A Sampler

on Sampling

WITHDRAWN

A Sampler
on Sampling

BILL WILLIAMS

Bell Laboratories
Murray Hill, New Jersey

John Wiley & Sons, New York • Chichester • Brisbane • Toronto

Published by John Wiley & Sons, Inc.

Library of Congress Cataloging in Publication Data:

Williams, William Howard, 1931-
   A sampler on sampling.

   (Wiley series in probability and mathematical statistics)
   Includes index.
   1. Sampling (Statistics) I. Title.

QA276.6.W54   1977   519.5'2   77-23839
ISBN 0-471-03036-8

Printed in the United States of America

10 9 8 7 6 5 4 3 2

To the family of my boyhood

and

to the family of my manhood

PREFACE

My purpose is to convey an *understanding* of the principles of statistical sampling to the nonmathematically inclined person. I believe that these principles are so important in our society that the educated person cannot remain ignorant of them and their consequences.

The concepts and procedures are presented almost entirely in the form of numerical examples from which generalizations are easily made. There is very little use of mathematical symbols, formulas, or development beyond the minimum necessary to follow the specific examples. Thus a background of about high school mathematics is all that is necessary to read this book. This includes some familiarity with summation notation, inequalities, simple equations, functional notation and elementary combinatorial analysis. Some of the exercises do require elementary algebraic manipulation. A summary of the notation used appears at the end of the book, after the subject index.

I have taught sampling many times in various educational programs. These include the University of California at Berkeley, the University of Michigan, Iowa State University, Rutgers, Stevens Institute of Technology and INCEP, which is Bell Laboratories in-house training program. This book has been heavily influenced by these various contacts. In particular, I have found that many people respond more readily to specific numerical examples than to the mathematics of sampling, which has an unfortunate characteristic of sometimes being both cumbersome and boring. As a result, I found that over the years, I was leaning more and more on numerical presentations. This book is a logical outgrowth.

The approach through numerical examples means that fewer ideas can be covered than in the typical text book on sampling. This is the price that must be paid for abandoning mathematical conciseness of notation, but for the average person the gain in readability seems worth the cost.

The loss is not as much as it might first appear, since there is only one basic idea in the whole book - the concept of the sampling distribution. Everything in the book is related directly to sampling distributions, how they arise, how they are related to the confidence placed in the sample, and how they can be influenced by such things as sample size, method of calculation of estimates, and sampling procedures. As the title indicates, no attempt has been made to be exhaustive, either in theory or methods. For example, I have barely scratched the surface of available sampling designs, and have said almost nothing about subsequent analysis of survey data (analytic studies). The selection has been based on my judgement of the ideas necessary for an understanding of the subject. Important new ideas are listed at the end of each chapter. Also, some representative books for additional reading in both sampling theory and practice are listed at the end of this preface.

This book is *not* intended to be a nonmathematical cookbook, nor should the reader be misled into thinking that the numerical example approach will make the basic sampling principles easier to understand. Ideas such as the manipulation of the sampling distribution seem to be hard to grasp regardless of how they are approached. In many ways, the numerical approach taken is similar to a mathematical one. The numerical examples are closely interwoven and follow a logical sequence. As a result, if the point of one example is missed, the next one may be difficult to follow, just as it would be if the approach were mathematical. Chapters 10, 11, and 12, particularly, need to be read carefully.

I do hope that the nonmathematical packaging of these ideas will make them accessible to many people who have neither a background in mathematics nor an inclination for it. The vastly increased use of sampling in government, industry, and in many academic areas of research, has greatly increased the number of persons needing an awareness and understanding of sampling. These people must use sampling to get the information they want, and need, as a basis for their research and decision processes. Hopefully this book will remove much of the mystery and black-box atmosphere of this vital step in their work. It is mainly for them that I have written this book, but it is also for the educated layman who wishes to follow intelligently what is being done and why.

Finally, since there are no mathematical proofs for the reader to study, I have in mind, as an operational suggestion, that the numerical results presented be verified, on a sample basis of course. Such spot verification will ensure that the development of the stated numerical results is understood and at the same time will prevent the reader from being discouraged by the rather formidable appearance of some of the tables. If this suggestion is carried out, a good understanding of sampling principles will be hard to avoid!

No book is ever written without substantial "outside" help. This book is no exception. First I must give considerable thanks to Bell Laboratories for many reasons of support, both direct and indirect. Perhaps, the most important is Bell Laboratories' persistent pursuit of applied statistics. This pursuit goes back to the late Walter Shewhart and is currently led by the very influential John Tukey. It is a unique organization.

Many individuals have contributed to this book. In particular, Colin Mallows, Dick Hamming, Henry Pollak, Roger Pinkham, Hwei Chen, Bob Brousseau, and Don Johnson read the manuscript in detail and suggested many improvements. Ram Gnanadesikan gave the necessary professional encouragement and Hwei Chen kept track of many numerical examples. Marvin Zelen pointed out the phenomenon described in the example of Section 2.1. Many students in my sampling courses also read the notes carefully. I thank them collectively. Above all, special thanks must go to Dick Hamming and Colin Mallows. These two associates were invaluable. I am greatly indebted to both of them. Finally, I wish to thank Penny Blaine for the many hours she spent typing, preparing, and phototypesetting the various drafts of this book. Her contribution was very substantial.

Over the years, I have benefited from personal interaction with many professional statisticians. From some of them I have learned valuable specific lessons. Consequently, I should like to thank Barbara Bailar, David Brillinger, Tore Dalenius, Ed Deming, Ram Gnanadesikan, Morris Hanson, Lou Harris, "HO" Hartley, Jon Rao, John Tukey, and Martin Wilk. I appreciate the discussions with all of them.

Murray Hill, New Jersey   Bill Williams
October, 1977

Additional Reading

Cochran, W. G., Sampling Techniques, 3rd ed. John Wiley & Sons Inc., New York, 1977.

Deming, W. E., Some Theory of Sampling, John Wiley and Sons, Inc., New York, N. Y., 1950.

Deming, W. E., Sample Design in Business Research, John Wiley & Sons, Inc., New York, N. Y., 1960.

Hansen, M. H., Hurwitz, W. N., and Madow, W. G., Sample Survey Methods and Theory, Vols. I and II. John Wiley and Sons, Inc., New York, N. Y., 1953.

Kish, Leslie, Survey Sampling, John Wiley and Sons, Inc., New York, N. Y., 1965.

Konijn, H. S., Statistical Theory of Sample Survey Design and Analysis. American Elsevier Publishing Company, Inc., New York, N. Y., 1973.

Lansing, John, B., and Morgan, James, N., Economic Survey Methods. Published by the Survey Research Center, The University of Michigan, Ann Arbor, Mich., 1971.

Yamane, Taro, Elementary Sampling Theory, Prentice Hall, Englewood Cliffs, N. J., 1967.

Yates, Frank, Sampling Methods for Censuses and Surveys, 3rd ed., Hafner Publishing Co., New York, N. Y., 1960.

CONTENTS

A Sampler

on Sampling

CHAPTER 1

CONFUSION

Errors of many kinds are made in sampling studies. In this chapter, examples are presented in which the source of confusion has been fairly well isolated. These examples contrast with those in Chapter 2, in which the source of difficulty has *not* been agreed upon. The examples in both chapters are also intended to stimulate interest in sampling, but they do not exhaust the apparently limitless number of possible errors. The achievement of good samples requires that we be as familiar as possible with potential pitfalls.

The examples illustrate different types of errors. "The Weather" (1.1) demonstrates that it must be clearly understood when and how all measurements are made. "Discrimination?" (1.2) shows that misleading conclusions can be reached even with correct data, and the "Draft Lottery" (1.3) points out the need for proper randomization. "The Club Meeting" (1.4) and "The Disappearing Children" (1.7) show how easy it is to bias sampling in favor of a special group, or certain individuals. "The Treatment Stack" (1.5) describes a transparent example of the subtle point that measurement procedures can interact with the study process itself so that completely misleading results are obtained. Finally, "The Aging Population" (1.6) shows that a sample which is useful for estimation at one point in time is not necessarily a useful sample at a later point in time.

Unlike the examples in Chapter 1, the examples in Chapter 2 do not have an agreed upon source of error. This is true even though the examples may appear to be very similar to some in this chapter. You are invited to formulate your own opinion whether the examples in Chapter 2 constitute "More Confusion?".

1.1 The Weather

A fancy travel brochure states that San Diego has 360 days of sunshine every year. Now the psychological appeal of such weather is so strong that it can preclude logical evaluation. It certainly did in my case; until I experienced 6 days of fog there I was not even suspicious!

1

While bad weather is always a vacation risk, the statistic of 360 days of sunshine out of 365 certainly suggests that 6 days of fog is an unlikely event indeed. What went wrong?

First, we suppose that the 360/365 statistic is *not* simply an outright lie. Then, the question becomes where and when were the measurements taken that led to the 360/365 statistic? Of course, the real answer to this question is not easy to track down, but for our purposes it is very informative to speculate about the source of the measurements.

San Diego often has morning fog near the ocean, but it almost always burns off before noon and rarely penetrates very far inland. So clearly, the time of day and location of the measurements will have a substantial effect on the results. And surely, the producer of an advertising brochure will use observations taken inland and *not* ones taken near the ocean!

But even after the variations due to time and location are considered, the statistic 360/365 can not yet be properly understood, because we do not know what a "sunny" day is. Is it a day with any sunshine at all? Or is it a day with a minimum of 8 hours of sunshine? The definition used for the brochure was not specified, but it really doesn't matter to us here because it is now clear that a brochure reader cannot have any precise idea of the meaning of the 360/365 statistic. There is just not enough information given to interpret the statistic in a usable way.

There is still one last lesson that can be taken from this example; it is that the observations are going to vary naturally, *even* if all the measurement and definitional problems are clarified. Amounts of sunshine will vary from one day to the next and from one year to the next, even in the same location. A year with no fog *may* be followed by a year with 12 days of fog.

So the answer to the earlier question of "What went wrong?" is that the statistic 360/365 was simply misinterpreted. Whether the data are from statistical samples or not, making the correct interpretation is a critical part of any study. It is an important issue quite apart even from the representativeness of the sample.

Exercises

1.1.1 Formulate a workable definition of a sunny day when,
(a) the weather is observed at a single location at 12 noon each day,
(b) the weather is observed hourly at a single location,
(c) the weather is observed hourly at a number of different locations in metropolitan San Diego.

1.1.2 Assume that all measurements are available to you without sampling. If you like, assume that you are "Nature" and know all about the weather. How would you summarize these measurements to represent the weather in San Diego?

1.1.3 In the 1976 Presidential election, President Ford seems to have fared somewhat better in states that he did not emphasize in his campaign than ones that he did. Should he have stayed at home?

1.1.4 An auto manufacturer advertises that 90% of its vehicles sold are still on the road. This is offered as a claim of durability. What do you think of this claim? Is your conclusion the same if most of the vehicles sold by this manufacturer were sold within the last 2 years?

1.2 Discrimination?

A major university was accused of discrimination against women.[1] The case seemed clear. An equal number of 100 men and 100 women[2] had applied to the university for admission and 46 men had been admitted but only 24 women. Since there was no evidence to argue that the men were better qualified than the women, the university did indeed seem to be guilty.

The question now arises as to who, within the university, is guilty. So the areas of Engineering, Science, and the Humanities were looked at separately. Historically, the *overall* acceptance ratios in these three

1. Bickel, P.J., Hammell, E.A., and O'Connell, J.W., Sex Bias in Graduate Admission: Data from Berkeley. Science, Vol. 187, No. 4175, 7 Feb. 1975, pp. 398-404.
2. For ease in illustration, a convenient set of artificial numbers has been used.

areas were 6/10, 4/10, and 2/10; so for example, in Engineering about 6 of every 10 students who applied were accepted. The acceptance ratios were largely a result of the availability of facilities relative to the number of students applying.

For the year in question, a total of 50 men and 5 women applied for admission to the various Engineering programs. However, the Engineering admissions board was careful not to discriminate against women and applied the overall acceptance ratio of 6/10 equally to *both* males and females. Consequently, 30 men and 3 women were admitted.

Science attracts somewhat more women than Engineering, and 30 men and 10 women applied. The Science administration was also careful not to discriminate between men and women and they too applied their overall acceptance ratio equally to men and women. The result was that 12 men and 4 women were admitted.

Many women apply to study the Humanities; this area is a well-documented favorite of theirs. Consequently, 20 men and 85 women applied. Again the admissions committee was careful not to discriminate against women and applied their overall admission rate of 2/10 equally to men and women. The result was that 4 men and 17 women were admitted.

Now an apparent contradiction exists. Each department *separately* acted fairly, but the overall result was that 46 of 100 men were admitted and only 24 of 100 women. So while the university as a whole appears to be guilty of discrimination, everyone within the university appears to be innocent!

The crux of the problem is that most women applied to the area of the university with the highest rejection rate. In fact, if the numbers of students applying to Engineering and the Humanities were interchanged, then 28 men and 56 women would have been admitted, just the reverse situation.

Initially, the university seemed clearly to be guilty of discrimination against women. But this conclusion is not necessarily correct because the aggregated data misled us. The general lesson from this example is not one relating to social equality but is rather that extreme care that must be taken in the interpretation of data - even accurate data.

Exercises

1.2.1   If a Federal grant is spent improving the position of female students in the University, one way might be to increase the facilities in the Humanities. This would enable the acceptance of more students in the Humanities. If the budget in the Humanities is directly proportional to the number of students how much bigger must this budget be in order that the *overall* University acceptance of males and females be equal?

1.2.2   A second suggestion to improve the position of women in the University is to give women preference in Engineering. However, it turns out that there is no money available to increase the Engineering facilities. As a result, the combined acceptance ratio for males and females together must remain at 6 in 10, that is a total of 33 students. If females are indeed given preference in Engineering, is it possible to equate the overall university acceptance of males and females by this procedure?

1.2.3   If women are encouraged to enter Engineering with the result that 50 men and 50 women apply, what will the overall university male/female acceptance ratio be? (Assume that the acceptance ratios for each of the areas are applied equally to both males and females.)

1.3 The 1970 Draft Lottery

In the late 1960s there was considerable controversy about the operational fairness of the draft laws. The procedure was largely controlled by local boards and many serious accusations were made. As a result, it was decided that for the 1970 draft the eligible candidates would be randomly ordered for induction.

To implement this random ordering, capsules representing each of the 366 days of the year were placed in a cage and randomly withdrawn by an individual. September 14 was selected first, which meant that eligible men born on that day would be the first ones inducted into the Army in 1970. The results of the entire drawing appear in Table 1.1.

A casual look at Table 1.1 does not suggest anything other than that the procedure was successful. However, it was not long before bitter complaints arose that the ordering was anything but random. The

TABLE 1.1

A DRAFTEE'S CALENDAR FOR 1970

Day   Month

Day	J	F	M	A	M	J	J	A	S	O	N	D
1	305	8	108	32	330	249	93	111	225	359	19	129
2	159	144	29	271	298	228	350	45	161	125	34	328
3	251	297	267	83	40	301	115	261	49	244	348	157
4	215	210	275	81	276	20	279	145	232	202	260	165
5	101	214	293	269	364	28	188	54	82	24	310	56
6	224	347	139	253	155	110	327	114	6	87	76	10
7	305	91	122	147	35	85	50	168	8	234	51	12
8	199	181	213	312	321	366	13	48	184	283	97	105
9	194	338	317	219	197	335	277	100	263	342	80	43
10	325	216	323	218	65	206	284	21	71	220	282	41
11	329	150	136	14	37	134	248	324	158	237	46	39
12	221	68	300	346	133	272	15	142	242	72	66	314
13	318	152	259	124	295	69	42	307	175	138	126	163
14	238	4	354	231	178	356	331	198	1	294	127	26
15	17	89	169	273	130	180	322	102	113	171	131	320
16	121	212	166	148	55	274	120	44	207	254	96	96
17	235	89	33	260	112	73	98	154	255	288	143	304
18	140	292	332	90	278	341	190	141	246	5	146	128
19	58	25	200	336	75	104	227	311	177	241	203	24
20	280	302	239	345	183	360	187	344	63	192	185	135
21	186	363	334	62	250	60	27	291	204	243	156	70
22	337	290	265	316	326	247	153	339	160	117	9	53
23	118	57	256	252	319	109	172	116	119	201	182	162
24	59	236	258	2	31	358	23	36	195	196	230	95
25	52	179	343	351	361	137	67	286	149	176	132	84
26	92	365	170	340	357	22	303	245	18	7	309	173
27	355	225	268	74	296	64	289	352	233	264	47	78
28	77	299	223	262	308	222	88	167	257	94	281	123
29	349	285	362	191	226	353	270	61	151	229	99	16
30	164		217	208	103	209	287	333	315	38	174	4
31	211		30		313		193	11		79	100	

problem is illustrated in Figure 1.1.

In Figure 1.1, the middle values, the medians, of the random assignments are plotted separately for each month. For example, in the January column of Table 1.1 15 numbers are above 211 and 15 are below. So, the median for January is 211, which was randomly associated with January 31. Clearly the later months have lower median values, which implies that men with birthdays in those months tend to be drafted earlier than men with birthdays earlier in the year. If the drawing were completely random, the medians plotted across months would appear to be level instead of having the declining trend which actually occurs.

What happened? While it can never be proven absolutely after the fact, it would appear that the capsules were put into the cage in monthly order and then not thoroughly mixed. As a result, the capsules tended to be drawn out together in monthly groups. This is an important lesson because, as we shall see later in this book, most useful statistical procedures are based on the premise of a proper randomization. Consequently, it must be carried out properly. This is not always easy to do in practice.

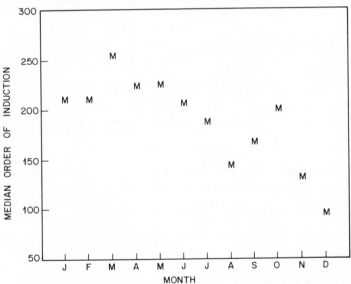

Figure 1.1. Systematic bias in the 1970 draft.

Exercises

1.3.1   The average ranking in Table 1.1 is $(1+2+\cdots+366)/366 = 183.5$. If these rankings are in fact only randomly related to birthdays, the monthly medians, plotted in Figure 1.1, should be randomly above or below the horizontal line with intercept 183.5. Draw this line with a ruler on Figure 1.1. Next draw a straight line that seems to you to pass closest to all of the medians plotted in Figure 1.1. Do you think the comparison of these two lines provides sufficient evidence that something went wrong with the mixing of the capsules? Explain your answer.

1.3.2   A list of "random" digits is a list in which the sequence of digits has been generated by some process which sequentially gives the integers 0, 1, 2, 3, 4, 5, 6, 7, 8, 9 an equal chance (1/10) of appearing. The table below contains 100 integers which are claimed to be random. What do you think?

03578	14689	23579	01457
02357	23689	12469	01468
83478	23679	81346	25789
25678	81349	12569	81458
84589	82789	13456	12367

1.3.3   The numbers 0, 1, 2, 3, 4, 5, 6, 7, 8, 9, 10, 11, 12, 13, 14, 15 can be represented in binary form as

$$n = c_1 2^0 + c_2 2^1 + c_3 2^2 + c_4 2^3$$

or more simply (c_1, c_2, c_3, c_4), where

$$c_1, c_2, c_3, c_4 = 1 \ or \ 0.$$

So for example,

$$0 = 0\times1 + 0\times2 + 0\times2^2 + 0\times2^3$$
$$7 = 1\times1 + 1\times2 + 1\times2^2 + 0\times2^3$$
$$15 = 1\times1 + 1\times2 + 1\times2^2 + 1\times2^3 .$$

Consequently, in binary form 0 is represented by (0,0,0,0), 7 by (1,1,1,0) and 15 by (1,1,1,1).

Select four coins, a penny, nickel, dime and a quarter and let c_i, $i = 1,2,3,4$ correspond to the penny, nickel, dime, and quarter respectively, such that

$$c_i = 1 \text{ if } a \ \ head$$
$$= 0 \text{ if } a \ \ tail.$$

For example, 0 corresponds to a tail on each of the four coins, 15 corresponds to a head on each of the four coins and 7 corresponds to a head on the penny, nickel and dime, and a tail on the quarter.

Now toss all four coins. Since the chance of a head or tail on each of the four coins is equal, we expect the numbers 0 through 15 to appear about the same number of times over repeated tosses of the four coins. Furthermore, if we ignore the numbers 10 through 15 when they occur we also expect the numbers 0 through 9 to appear with approximately equal frequency.

Using the process described above, generate 100 random digits 0 through 9. Then evaluate the randomness of your set of numbers by the same criteria that you used in Exercise 1.3.2.

1.3.4   To overcome the difficulties associated with the 1970 draft, the method of randomization for the 1971 draft was different. Two containers of capsules were used. One set of capsules in one container represented the 365 birthdays in 1971 and the other capsules in the other container represented the integers from 1 to 365. Then the capsules were mixed and drawn in pairs, one from each container. In this way, January 10 was paired with the integer 101, and July 9 was paired with the integer 1. This meant that men born on July 9 would be drafted first and that men born on January 10 would be drafted only after those men born on days associated with the first 100 integers. The complete results are given below. Use any analysis you like to decide how successful the randomization was in 1971.

A DRAFTEE'S CALENDAR FOR 1971

Day   Month

	J	F	M	A	M	J	J	A	S	O	N	D
1	133	335	14	224	179	65	104	326	283	306	243	347
2	195	354	77	216	96	304	322	102	161	191	205	321
3	336	186	207	297	171	135	30	279	183	134	294	110
4	99	94	117	37	240	42	59	300	231	266	39	305
5	33	97	299	124	301	233	287	64	295	166	286	27
6	285	16	296	312	268	153	164	251	21	78	245	198
7	159	25	141	142	29	169	365	263	265	131	72	162
8	116	127	79	267	105	7	106	49	108	45	119	323
9	53	187	278	223	357	352	1	125	313	302	176	114
10	101	46	150	165	146	76	158	359	130	160	63	204
11	144	227	317	178	243	355	174	230	288	84	123	73
12	152	262	24	89	210	51	257	320	314	70	255	19
13	330	13	241	143	353	342	349	58	238	92	272	151
14	71	260	12	202	40	363	156	103	247	115	11	348
15	75	201	157	182	344	276	273	270	291	310	362	87
16	136	334	258	31	175	229	284	329	139	34	197	41
17	54	345	220	264	212	289	341	343	200	290	6	315
18	185	337	319	138	180	214	90	109	333	340	280	208
19	188	331	189	62	155	163	316	83	228	74	252	249
20	211	20	170	118	242	43	120	69	261	196	98	218
21	129	213	246	8	225	113	356	50	68	5	35	181
22	132	271	269	256	199	307	282	250	88	36	253	194
23	48	351	281	292	222	44	172	10	206	339	193	219
24	177	226	203	244	22	236	360	274	237	149	81	2
25	57	325	298	328	26	327	3	364	107	17	23	361
26	140	86	121	137	148	308	47	91	93	184	52	80
27	173	66	254	235	122	55	85	232	338	318	168	239
28	346	234	95	82	9	215	190	248	309	28	324	128
29	277		147	111	61	154	4	32	303	259	100	145
30	112		56	358	209	217	15	167	18	332	67	192
31	60		38		350		221	275		311		126

1.4 The Club Meeting

Not so long ago, I belonged to an organization that met twice a month, always on a Tuesday or Thursday. There was no particular regularity to whether the meeting was held on Tuesday or Thursday; it was more that the meeting should not be held on a Monday, Wednesday, or a Friday because many academic members had heavy teaching schedules on those days. Attendance was typically about 25% of the membership, and it was felt that this could be improved if the meetings were always held on the same week day. People could then plan ahead better. So, a poll was taken. It was taken at a Thursday meeting, but with no proposal to repeat the process at a subsequent Tuesday meeting. Thursday won hands down. The "can-only-come-Tuesday" people never had a chance. Further compounding this outrageous bias was the fact that the chairman was a statistician. He had clearly "stacked the deck" in favor of Thursday and created a beautiful illustration of poor statistical procedure.

Other examples of biases created by stacking the deck are discussed later in the book.

Exercises

1.4.1   The club officers decided to conduct the day-preference poll at both a Tuesday and a Thursday meeting. At the Tuesday meeting, the vote was 3:1 in favor of Tuesday but at the Thursday meeting the vote was only 2:1 in favor of Thursday. As a result the club officers were about to vote for all Tuesday meetings until it was pointed out that they might be making another mistake. What was it? The attendance was 100 at the Tuesday meeting and 150 on Thursday.

1.5 The "Treatment" Stack

In Bell System business offices, it is a major responsibility of the Service Representatives to contact customers who have not paid their bills. This function is called "treating" the customers.

An interesting situation arose when an estimate was obtained that 40% of the Representatives' time was spent "treating" customers; prior estimates had suggested a maximum of 20%. So the sampling study which yielded the 40% estimate was scrutinized.

At random times of the day, an observer went to the desk of each Service Representative to note the current task, but the observer *always* observed the Representatives in the *same* predetermined order! This meant that as soon as the observer appeared on the floor each Representative knew that an observation was about to be made and had enough time to change tasks. The effect was that if a Service Representative was doing a "less desirable" (but still legitimate) task such as waiting for the next incoming call, there was a very strong tendency to pick up the "treatment" stack of bills (which was always there) and start treating customers. Apparently, this tendency was strong enough to create a substantial overestimate of "treatment" time.

This bias was eventually reduced by the use of independent observation times for *each* of the Representatives. In this way, no one knew in advance when an observation was to be made, and the distortion resulting from the interaction of the measurement process and the persons being observed was avoided.

The clear lesson is that the measurement scheme should not distort the process that is being measured. Such distortion arises particularly easily with human populations, and is certainly possible even with the "hard" observations of the physical and biological sciences.

1.6 The Aging Population

Samples of people in the United States are often selected for repeated observation through time. Such a sample is called a *panel* survey and if the sample has *no* rotation of persons in and out of the sample, it is called a *fixed* panel survey. The reason usually given for such surveys is that "tracking permits the maximum assessment of changes through time." Consequently a fixed panel may be used by the telephone company to assess the effects of a rate change, or by the government to assess the effect of a change in taxation on consumer spending.

On the surface of it, and even from some statistical points of view, a fixed panel survey sounds very reasonable. But suppose that one of the items of interest is the age of the United States population. Clearly, the average age of the fixed panel increases by one year every year; but the average age of the United States population certainly does *not* increase that fast, and indeed, if the birth rate is high, the average age of the population may actually fall! As a result, estimates of population age cannot be derived from a fixed panel, except at the original time of selection. This is true even if, on the first day of the study, the panel consists of every living resident of the United States. Death rates can then be observed but birth rates cannot. Furthermore, since the panel ages one year every year, estimates of *change* made from this panel are also very poor. All of this suggests that anything highly associated with age such as health and income will also be treated poorly. The general argument "that tracking permits the maximum assessment of changes through time" can be badly misleading.

The problem is that any similarity between sample and population may disappear rapidly after the initial sample selection. This can happen either because the characteristics of the sample change more rapidly than the population (as in the age example), or because the population changes more rapidly than the sample. In any event, the onus is on the sampler to justify the later use of a fixed panel sample, because representativeness of a sample at one point in time does not automatically justify its use at a later time.

Panel surveys are extremely common. To name a few areas, they are used in demographic, engineering, marketing, and medical studies. But unfortunately, many panel biases are a good deal more subtle than the one exhibited by the age example. It is clear that they must be used with care.

Exercises

1.6.1   Fixed panel samples are never really quite fixed. That is, even if the sample design calls for the original sample to be reinterviewed in its entirety some sample units will be lost from the study.

In a marketing study, customers of a major United States utility were tracked through time in a fixed panel sample. For various reasons, some customers were lost from the sample. Two such reasons were physical death of the customer and moving out of the geographical sample region. At the end of a period of time, say 5 years, do you think the remaining customers could be used as a representative sample of the company's customers *at that point in time?* Do you think those remaining customers are representative of customers at the original time of selection? State reasons for your answers. If your answer is "maybe" state the circumstances under which the yes and/or no answers are appropriate.

1.6.2   This exercise refers to Exercise 1.6.1. In an effort to keep the sample size "up," the panel of customers was replenished by a selection of customers who were entirely new to the company. These new customers would replace those "old" customers who had been lost from the sample. In this way, the planners of the study hoped to keep the sample representation balanced. Discuss the pros and cons of this balancing.

1.7 The Case of the Disappearing Children

A marketing study in a midwestern city produced data that clearly indicated a rapid and significant disappearance of children. Groups of city blocks were drawn into the sample and the persons living there were interviewed for three consecutive months. The first month that the areas were canvassed the average number of children per family was 3.2, followed by 2.5 in the second month, and 2.4 in the third. Now while the 2.5 and 2.4 averages of the second and third months are not very different, the 3.2 average certainly is. Where were these children going? Surely the parents were not that forgetful! It turns out that there is a simple and logical explanation. To understand it, we must look at the sampling procedures more closely.

The sample design specified that a subset of households on the sampled city blocks be interviewed. Not surprisingly, not all the selected households were actually interviewed, for the simple reason that not everyone was at home. In practice, all household studies have such response problems; only the magnitude of the nonresponse varies.

Procedures are usually implemented to minimize nonresponse. For example, interviewers are trained to ask respondents about the habits of their not-at-home neighbors so that the chance of getting a response in the later months can be maximized. As a result, most panel surveys experience increases in the overall response rate the longer the areas are retained in the survey. This was in fact true of this marketing study.

Who then are the nonrespondents? Early in the study, the answer is families with no children. Such families are away from home far more than families with children and as a result are less likely to be interviewed. As the survey progresses, the operational procedures described previously increase the chances of obtaining interviews from the originally nonrespondent families. The result is that initially families with children are substantially over-represented, but as the survey progresses they are more properly represented, with the total result that there is a systematic "loss" in the number of children.

The lesson of this example is an important one, namely that samples are very easily biased. Compounding this problem is the fact that the possibility of bias is easy to ignore. Biases are discussed frequently in this book.

Exercises

1.7.1   A sample of 100 adults is selected to study adult weights. The mean weight of the sample is 150 lb. Unfortunately, it is discovered that males seem to be over-represented in the sample; specifically, there are 75 males and 25 females. Since males and females are about equally represented in the population, it would appear that the mean weight of 150 lb. is too high to represent the population at large. If the mean weight of the sample males alone is 160 lb. and of the sample females 120 lb., how would you construct a new estimate of weight which adjusts for the over-representation of males?

1.7.2   In view of the method that you used in Exercise 1.7.1, what information would you seek to adjust the estimates in the survey described in Section 1.7? How would you use this information?

CHAPTER 2

MORE CONFUSION?

In Chapter 1 some errors and misinterpretations of survey data were discussed. Those examples are now generally thought to be pretty well understood. In this chapter, examples are discussed which have some of the flavor and apparent difficulties of those of Chapter 1, but for which the sources of the difficulties are still unresolved. The examples are described only briefly; many of the details are omitted for the sake of getting rapidly to the key points. See whether you can decide yourself, on the basis of these descriptions, whether confusion exists or not.

2.1 Uterine Cancer

During the 1950s the use of the Pap test for the detection of uterine cancer spread rapidly and was thought to have a major reducing effect on the number of female deaths from this type of cancer. And indeed, uterine cancer deaths in the United States dropped steadily from about 18 per 100,000 females in 1950 to about 11 in 1970. The decline is often quoted as proof of the effectiveness of the Pap test. The conclusion seems to be entirely reasonable.

Figure 2.1 shows the death rates from uterine cancer from 1930 to 1970. The data source is the National Vital Statistics Division of the Bureau of the Census. The graph clearly shows that deaths from uterine cancer began dropping *well before* the widespread use of the Pap test! This immediately raises the question of whether the drop in uterine cancer deaths could have an entirely different cause. Is it possible that, just like the weather in Section 1.1, this is an example of the interpretation of incompletely understood data? To be sure, the cancer question is far more complex than the weather illustration; nevertheless, the fact that deaths from uterine cancer began dropping well before the Pap test appears to be unexplained.

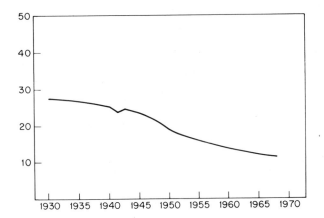

Figure 2.1 Deaths from uterine cancer
per 100,000 female population

Exercises

2.1.1   The American male is apparently badly clothed. In a given
year only 19% of all adult males buy a new suit. Discuss this
statement.

2.2 Sexual Activity

In the 1950s, the Kinsey report[1] was widely discussed as a source
of information about human sexual activity. The statistical summaries
used in the report were based on a population survey in which respondents were asked about their sexual behavior.

The question for discussion here is whether such information
should be regarded as a report of what people did, or of what they *said*
they did? It is easy to imagine many people overstating their real

---

1. Kinsey, Alfred C., Pomeroy, Wardell B., and Martin, Clyde, E., Sexual Behavior in
the Human Male. W. B. Saunders Company, Philadelphia, 1948.

activity. Other persons, no doubt, reacted in the opposite way. Consequently, is it possible that the method of measurement (the personal interview) ensures that accurate data cannot be obtained? Unfortunately, it is not easy to see how accurate answers to such sensitive questions can be obtained at all, although one suggestion is made in Section 6.3, "The Sampler's Approach."

This difficulty is similar, but not quite the same as the "treatment" stack of Chapter 1. In that example, the measurement method caused a *real* change in activity. In the present case, real sexual activity probably did not change since the survey was retrospective, but there is a major concern that the description of sexual activity was distorted. Another difference is that it is not possible to check the accuracy of descriptions of sexual activity, while in the "treatment" stack study, taking the measurements in an entirely different way led to more realistic results.

This type of measurement difficulty exists for all surveys that depend upon human responses. Clearly, such responses describe what people *said* they did, or would do. But whether these responses also describe what *really* happened should be considered with great care for each measured item.

2.3 Retrospective Samples

Newspapers often describe the death dealing effects of pollutants in the environment; they state, for example, that a suspiciously large number of workers exposed to asbestos have died of lung cancer. Similar statements have been made about vinyl chloride and other chemicals.

What is the nature of the statistical studies upon which these statements are based? Presumably, such a study begins when a suspicious researcher reviews medical records counting cancer deaths among people who worked in a particular plant or who lived in a particular geographical location. Of course, to be meaningful such a count must be compared with the total number of workers exposed, which raises the question, "Is the number of exposed persons *not* contracting lung cancer counted as diligently as the number who do?" Clearly, if workers who remain cancer free do not have the same chance of being counted as those who do develop this disease, an overestimate of cancer incidence will result; just as the numbers of children were overestimated in the example of Section 1.7. To be convincing, such retrospective studies must demonstrate that the counts of *both* diseased and healthy persons are made with equal diligence. Do you suppose that such equal diligence is really applied?

Exercises

2.3.1   During the August 30, 1976, 10 p.m. newscast, New York
          television channel 5 conducted a telephone poll on the ques-
          tion, "Should the government pay for college education?" Peo-
          ple who wished to vote yes were asked to call one telephone
          number and those who wished to vote no were given a
          different number. Presumably each of the telephones was set
          to count the number of incoming calls. It was announced that
          over 4000 yes votes had been received and over 5000 no
          votes. Unfortunately, it was later discovered that a technical
          difficulty had prevented some calls to the "no" telephone from
          being counted. Discuss the possible effects of this problem.
          In particular, would it affect any general conclusions that might
          have been made?

2.4 Joggers

The life expectancy of joggers is often compared with nonjoggers.
Tables comparing these two groups of people are easy to find and
invariably show that joggers live substantially longer. The reason usually given for this increased longevity is that the jogger develops a superior cardiovascular system. The conclusion? Begin jogging and develop
a good cardiovascular system!

The question here is, do we have the argument backwards? Is it
possible that joggers jog because they already have superior, or potentially superior, cardiovascular systems? If so, the joggers' longevity
could be the result of a poorly understood third factor which influences
both their jogging ability *and* their longevity.

This is a specific example of the general problem of resolving
which, if either, of two related variables is cause and which is effect.
Statistical arguments cannot answer such questions. They must ultimately be resolved by an understanding of the fundamental characteristics of the variables under study. In the case of jogging, there does

seem to be some underlying understanding of physiological reasons for the expectation that jogging and physical activity will indeed improve longevity.

This assignment of cause and effect is a classical problem; there are many other examples. Rum prices and ministers' salaries are known to be correlated, but does this mean that to increase ministers' salaries the price of rum should be increased? Or vice versa? In this case both factors tend to rise together as a result of inflation and are probably only very distantly related. Certainly, no useful cause-effect relationship is apparent. Unfortunately in general, analysis of cause and effect is often very difficult.

Exercises

2.4.1   It is known that in later life graduates of the Harvard Business School have incomes well above average. The recipe for success would then seem to be get yourself into the Harvard Business School by any means possible because this will guarantee a life-long high income. What do you think of this conclusion?

2.4.2   A company's sales increased steadily for 5 years. The advertising budget of this same company also increased steadily for the same 5 years. The advertising director argues that this proves the effectiveness of his program. Do you agree? What other explanations could there be? If you were company president what additional information would you ask for?

2.4.3   National Statistics show a strong relationship between the duration of a marriage and the number of children produced by that marriage. Clearly then if a couple wishes to have a long marriage they should have many children. Discuss.

2.5 Television Ratings

In "The Aging Population" (1.6) it was pointed out that severe bias may result from the repeated use of the same sample. Such samples are sometimes called fixed panels. The potential bias difficulties were apparent in the aging population because the persons in the sample were getting one year older every year, while the population as a whole was not. In general, discrepancies between the sample and population can arise either because the sample changes through time or because the population does.

Furthermore, fixed panels should not necessarily be used even for estimates of *change* in the population. Again, this was seen in the "Aging Population" example of Chapter 1. All in all, it should be clear that the use of a panel survey is a delicate operation which is not to be treated casually.

Some television ratings are based on panel surveys. Black boxes are attached to the respondents' television sets, so that they can be observed repeatedly through time. No doubt it would be a headache to reposition these measuring devices every time that new observations are needed. But keeping in mind the very serious bias problems of the age estimates, can we be sure that the television ratings are not similarly, if perhaps more subtly, biased? Unfortunately, few studies based on fixed panels pay much attention to the possibility of systematic bias.

Were you annoyed when your favorite television program was dropped? Perhaps you had good reason to be. Maybe the survey panel now has an average age over 30 - that terrible age at which all insight and judgement are lost!

2.6 Unemployment

Unemployment estimates in the United States are derived from the Current Population Survey (CPS) conducted by the United States Census Bureau. The CPS sampling design is complex, but for our purposes we need only to understand its rotating panel features. The CPS is conducted monthly, and each month, one-eighth of the sample is selected new. Each one-eighth group is retained in the sample for four consecutive months, dropped for the next eight months, after which it is brought back into the sample for four consecutive months. In this way each rotation group appears in the sample for a total of eight months.

This design is a modification of the fixed panel in that each month part of the sample is fixed and part is new. It is clearly a more complex design than the completely fixed panel design discussed in the aging example of Section 1.6. In addition, the CPS fixed "panels" are geo-graphical areas and all people residing in these areas are potentially included, whether or not they are new residents in the area. But, in spite of these differences the CPS is partially a fixed panel, conse-quently it may be open to the same kind of criticism as the aging and television examples. However, it is by no means obvious that it is.

In 1962, a Presidential[2] committee reported on a "first month" bias problem in the estimation of unemployment. The problem was that the area panels of the CPS consistently reported far higher unem-ployment the first month of observation than in the later months. As it turns out the problem can be described in a somewhat more general way than it was in the Presidential Report.

Table 2.1 shows reported unemployment versus the number of times in the survey. Unemployment *appears* to be highest for panels in the sample for the first time, falls off for the next 3 months, then rises at month 5 and falls off again. (Recall that there is an 8-month lapse between the fourth and fifth appearances.) These changes in the rate of unemployment are much larger than standard statistical norms would predict. Why then does this characteristic, which appears quite con-sistently, exist in the estimation of unemployment, and as well as some other measured items, such as income?

TABLE 2.1

CPS TOTAL UNEMPLOYMENT 1955-61

Appearance in Sample

	1	2	3	4	5	6	7	8
Index	107.3	100.3	100.3	98.9	100.7	99.6	96.6	95.0

(Index numbers, all groups combined equals 100)

In human surveys, not everyone can be interviewed. Non-response occurs because people are not at home, refuse to answer, or cannot be found. However, as the survey progresses and attempts are

2. "Measuring Employment and Unemployment" Report of the President's Committee to Appraise Employment and Unemployment Statistics, Washington, D. C., 1962.

made to reinterview the same persons, the response rate usually rises because interviewers are carefully instructed in methods to maximize response. This rise is in spite of a usual increase in the numbers of people who refuse to cooperate.

CPS response rates follow a pattern that is very similar to the unemployment rate. These response rates are given in Table 2.2 and it can be seen that they exhibit the same (although reversed) pattern as the unemployment rates, Table 2.1.

TABLE 2.2

CPS RESPONSE RATES 1955-61

Appearance in Sample

1	2	3	4	5	6	7	8
90.1	94.3	95.1	96.1	93.4	95.0	96.2	96.8

This immediately raises the question: Are the rates of response related to the peculiar behavior of the unemployment estimates?

First, there is a belief, with some documentation, that in a newly selected area unemployed persons are more likely to be found at home than employed persons. (Be careful, this says "likely to be found at home" *not* that there are *more* unemployed persons than employed ones.) Second, there is a hard core of unobservable, unemployed persons in central urban areas. It seems that these people cannot be counted or interviewed no matter how much effort is expended. In fact, shortly after the 1970 census, the then New York Mayor John Lindsay, called a special meeting to discuss just this problem. He felt that New York City had been under-represented in these core areas.

If the two factors described in the previous paragraph are in fact operative, a systematic bias in unemployment estimates of exactly the kind alluded to in the Presidential[2] report will result. In the first months of observation unemployed people tend to be over-represented and in the later months they tend to be under-represented. So, could it be that United States unemployment estimates have selection bias difficulties of the same kind, although more complicated, as the Disappearing Children of Section 1.7?

Finally, we add that it has been shown[3] mathematically that even very slight distortions in the chance of response by any identifiable group, such as unemployed persons, can indeed cause very major biases.

Exercises

The next few exercises are designed to illustrate some of the unfortunate effects of nonresponse. Ignore any difficulties of definition of unemployment and unwise sample selection.

2.6.1   Suppose that authorities wish to know the unemployment rate, R_u in a small town. Furthermore, suppose that to determine R_u they decide to interview everyone of the 10,000 people in the town's work force. Finally suppose that 4% or 400 of the work force is unemployed.

Unfortunately, not all 10,000 people will be interviewed; some will be out at work, on vacation, and so on. If 90% of the people who are employed and 95% of those who are unemployed are interviewed, calculate n_e the number of employed people who are actually counted and n_u the same number for those who are unemployed. Next calculate the observed unemployment rate $r_u = 100 n_u/(n_u + n_e)$. Discuss this observed rate, r_u, in comparison with the true rate $R_u = 4.00\%$. For example, why is the observed rate of unemployment too high?

2.6.2   Seven different cases of possible response rates are listed below. Case i is the same case as in Exercise 2.6.1. For the other six cases, verify that the observed unemployment rate is as given.

---

3. Williams, W. H., and Mallows, C. L., Systematic Biases in Panel Surveys Due to Differential Nonresponse. Journal of the American Statistical Association, Vol. 65, pp. 1338-1349, September 1970.

Case Number	Percent Interviewed		r_u
	Employed Persons	Unemployed Persons	
i	90	95	4.21
ii	95	95	4.00
iii	95	90	3.80
iv	95	98	4.12
v	96	90	3.76
vi	92	95	4.13
vii	98	95	3.88

2.6.3   Referring to Exercise 2.6.2, discuss the results of each of the cases ii-vii in the same way that you considered case i. Be sure to point out why r_u is higher or lower than the true value of 4.00 percent. Give special attention to case ii in which r_u is right on target.

2.6.4   In some studies people are interviewed repeatedly over a period of time. For example, in the unemployment study, a sample of the work force may be interviewed on consecutive months to estimate changes in the unemployment rate. If case i in Exercise 2.6.2 refers to month 1 and case vii refers to month 2 it would appear that the unemployment rate has dropped about 8% from 4.21 to 3.88, when in fact the true rate R_u is 4% on both occasions. Explain this phenomenon. Can you suggest ways to prevent being misled in this way?

2.6.5   In Exercise 2.6.2, it was necessary to calculate n_e and n_u, the *actually* counted numbers of employed and unemployed persons. The ratio $100(n_e+n_u)/10,000$ gives the overall response rate. Confirm that the overall response rates for the seven cases are as given here.

Case Number	i	ii	iii	iv	v	vi	vii
Response Rate	90.2	95.0	94.8	95.1	95.7	91.9	97.8

2.6.6   As household surveys go, the response rates calculated in Exercise 2.6.5 are very high. If the overall response rate was lower, say 50 to 60%, do you think the difference between the observed r_u and the true value of 4.00 might be even larger than the difference exhibited in Exercise 2.6.2? Construct some examples to illustrate your answer.

WHAT ARE YOU
TALKING ABOUT?

3.1. Target Populations

Samples are used to represent some larger group. In fact, the main interest lies in this larger group, which in statistical jargon is called *the target population*. Target populations can be very different. The entire United States adult population is frequently the target population for studies conducted by the various national opinion polls. In other surveys, the target population consists only of those persons living in a certain state, or city. Target populations do not have to consist of people. A day's production of transistors, or all telephone calls made during a day are also possible target populations. Except for operational difficulties, virtually any group can become the target population of a sampling study. The possibilities are endless.

Groups are made up of individual *elements.* The individual elements in the target populations are referred to as *population elements,* or sometimes as *population units.* For example, in the population of United States adults, the units are individual United States adults. It follows almost without saying that if there is to be a clear understanding of the target population, there must be a clear understanding of the population units. The definition of the population units must always tell which units are in, or out, of the target population. For example, in the study of United States adults, a clear definition of an adult is required. Does this mean over age 18? Over 21? It must also be clear what "United States" means. Does it mean United States citizens only? Or United States residents? Must they be in the country at the time of the study? Operationally, a sample of United States adults cannot be drawn without a precise definition. As we saw in the weather example (1.1), vagueness about these definitions makes interpretation of a survey very difficult, if not impossible.

Precise definition of the population units makes it *theoretically* possible, given enough resources, to list all units in the target population. For example, we could theoretically list every United States adult. In statistics, such a list is called *a frame*. Fortunately, as we shall see, it is nearly always unnecessary to actually construct the complete frame list. However, the precise definitions, which would permit construction of the list, are very necessary.

In practice, frame "lists" often carry more information than just an "in" or "out" definition. They also contain information about the grouping of the population units. For example, in the study of all United States urban adults, the frame list might include the state and city of each person's residence. For city residents, the city block and street address may also be included. In this illustration, there is a hierarchal grouping of the population units which are grouped within blocks, the blocks are grouped within cities, and the cities are within states. In statistical terminology, the largest groups (states) are called primary groups, the next (cities) are called secondary groups, and the third (blocks[1]) tertiary groups.

The grouping of the United States adult target population described in the previous paragraph is a *natural* grouping of the population. An understanding of such grouping is necessary because as we see in Chapter 11, "The Clever Use of Groups," this understanding can be used to improve the efficiency of the sampling. In fact, grouping can be so useful that it sometimes pays to create *artificial* groups. On the other hand, a bad sampling method not only does not profitably utilize these groupings but actually permits them to create inefficiencies.

Exercises

3.1.1   To test for water purity, engineers wish to sample the water as it passes from storage reservoirs into city consumption. Before this can be done, however, a target population must be defined. Use your imagination to define such a population. Would you expect that water impurity might be seasonal? If so, should this knowledge affect the duration covered by the study? In what other ways might water impurity vary? Be sure to define your population units.

---

1. Blocks or groups of blocks are in fact often used in urban surveys.

One often hears complaints about how misleading various surveys are. The following problems are intended to show how difficult some apparently "easy" problems really are.

3.1.2   To estimate the rate of unemployment it is necessary to have a definition of an unemployed person. What is your definition of an unemployed person? Be sure that your definition answers the following questions. Is a student unemployed if he cannot find a job during the summer? Does the age of the student make a difference? Is an elementary school teacher unemployed during the summer? Does it make a difference if the teacher wants a job or not? Sometimes housewives take jobs and leave them as they become available to them and as family matters permit. When should such people be counted as unemployed? To be at all useful, your definition of an unemployed person must not only account for all of the preceding questions; it must also account for many that have not been asked. Be sure that your definition anticipates as many of these questions as possible.

3.1.3   To measure the unemployment *rate,* the number of persons unemployed must be compared with the number of people in the "work force." Try to formulate a precise definition of the work force.

3.1.4   Over the years the United States Department of Agriculture has carefully developed a definition of a farm. If you were faced with the task of studying United States farms how would you define a farm? Is the city executive with one acre of land operating a farm on weekends? How about the executive with 10 acres? A hundred? Must the produce be sold in order that an operation be classified as a farm? Consider as many such questions as you can.

3.1.5   Many samples are now drawn by random-digit telephone dialing. This means that the population consists of telephone customers with all possible 7-digit telephone numbers. Usually the goal of the survey is to draw inferences to the larger population of all United States households. Compare these two populations.

3.2. A Taxpayer Population

The headline in the Sunday, April 14, 1974 San Francisco Chronicle and Examiner said, "Billions in lost taxes ... and IRS can't cope." The article contained the data given in Table 3.1.

TABLE 3.1

ERRORS IN INDIVIDUAL TAX RETURNS

Individual Returns	Percent of Returns with Errors	Average Tax IRS Says is Owed but Not Paid (dollars)	Total Tax Owed but Not Paid (billions)
Standard Form 1040	32	220	1829
Income Under 10,000 dollars deductions itemized	49	178	1821
Income between 10,000 dollars and 49,999 dollars	57	303	3522
Income 50,000 dollars and over	82	8631	2236

There is much interesting information in Table 3.1, but for us the table leads to a sequence of questions. The first question is: What is meant by an "error"? For our purposes, it is assumed that an "error" is a difference between the reported taxable income (call it x) and the actual taxable income (call it y), that is, errors are the differences $(y - x)$. The second question is: How was such information obtained? Now determining the actual income (y) implies a special IRS audit and since audits are expensive, the table is probably based on a sample of taxpayers. How then was such a sample drawn and analyzed? Since the article does not say, and since all of us are involved in the IRS process, the question makes a natural background for presentation and discussion of the techniques of sampling.

To make the discussion concrete, a small hypothetical target population of nine taxpayers has been invented. The actual and reported incomes are given in Table 3.2. Just as a real study of taxpayers would presumably focus on persons with large incomes, the numbers in Table 3.2 are large.

But the *size* of the artificial taxpayer population is small, $N=9$. This has both advantages and disadvantages, but for our purposes the advantages outweigh the disadvantages. Real target populations are often very large and may contain thousands, or even millions of population elements. This makes presentation difficult and visible manipulation impossible. On the other hand, a small population can be manipulated and discussed easily. Furthermore, an artificial example has the advantage of not introducing any of the operationally difficult factors which often surround real samples and which can cloud the points under discussion.

In any event, sampling principles are the same whether the target population is large or small. But this is not to say that size differences are not important in practice. They clearly are, if only for the reason that there are many more opportunities for foul-ups in a large study than a small one.

TABLE 3.2

A TARGET POPULATION OF TAXPAYERS

Taxpayer	Actual Taxable Income (y) Thousands of Dollars	Reported Taxable Income (x) Thousands of Dollars
T_1	60	50
T_2	72	56
T_3	68	66
T_4	94	76
T_5	90	90
T_6	102	100
T_7	116	112
T_8	130	110
T_9	200	175
Total	$Y = 932.00$	$X = 835.00$
Mean	$\bar{Y} = 103.56$	$\bar{X} = 92.78$

$$\sigma_y^2 = 1621.14 \quad \sigma_x^2 = 1296.40$$

$$S_y^2 = 1823.78 \quad S_x^2 = 1458.44$$

In reality, the IRS will know the reported income (x) for each taxpayer, but the actual taxable income (y) will be known only for audited taxpayers. However, for discussion in this chapter, suppose that everything is known, that is, y and x are known for every taxpayer. What then are the interesting features of this taxpayer population? The purpose of this question is to determine what information we would like to obtain by the sampling procedures which we shall begin discussing in Chapter 5.

A plot of the actual incomes (y) of this target population appears in Figure 3.1. Income is plotted on the horizontal axis and the frequency of each of the observations is on the vertical axis. In this example, each different income figure occurs exactly once, and as a result the plot takes on a particularly simple appearance. Nevertheless, it does convey some information because it permits a visual study of the way in which the incomes are *distributed* along the income axis. In fact, a plot like Figure 3.1 is sometimes referred to as a plot of the income *distribution*.

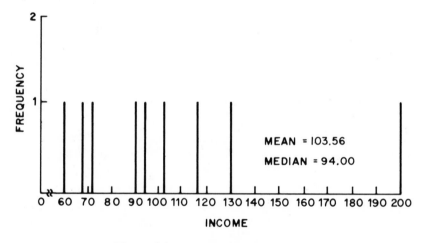

Figure 3.1 Actual taxpayer income.

First, notice that the average income is $\overline{Y} = 932/9 = 103.56$ (thousand) dollars which tends to locate a "*center*" or "*middle*" of the distribution of the individual incomes. This simple average, $\overline{Y}$, which is sometimes also called the mean, is by far the most popular measure of the location of the "middle" of a distribution.

The simple average is not completely satisfactory however. Notice that only three of the nine taxpayers have incomes greater than the mean 103.56. To a great extent this is a result of the one very large income of 200,000 dollars which increases the average income substantially. This is a general characteristic of the simple average; it is very much influenced by numbers that are different from the rest. In statistics, such numbers are sometimes called outliers.

As a result of this sensitive characteristic of the average, there is some interest in location measures that are not as easily influenced by a small percentage of the numbers. The median is one such location measure. If the number of observations is odd, the median is the middle one of these observations; if the number is even, the median is the average of the two innermost observations. So in the taxpayer population, the median income is $Y_{med} = 94.0$. Notice that if the one large taxpayer is not included, the median of the remaining eight values is $(90+94)/2 = 92$. This is a drop from 94 to 92. In contrast, eliminating the large taxpayer drops the mean $\overline{Y}$ from 103.56 to 91.50, which is a much larger drop than the median.

There are other measures of location, but the mean and median are the only two that we shall introduce now.

The spread of a distribution is also of interest. Certainly, there has been substantial political interest in the variability of individual incomes. But variability is present in all relevant target populations; after all, if there is no variability in the numbers, there is little reason to study them.

There are various ways to measure variability, the most obvious of which is the *range* of the observations, that is, the largest observation minus the smallest. In the taxpayer population, the range R_Y, is readily observed to be equal to $200-60=140$.

But while the range may be the most obvious spread measure, it is by no means the most used. This distinction is reserved for a quantity called the *variance* and its square root, the *standard deviation*. The variance is calculated as the average of the sum of squares of the deviations from the mean. So, for the taxpayer population,

$$\sigma_y^2 = \frac{1}{9} [(60-103.56)^2+(72-103.56)^2+ \cdots +(200-103.56)^2]$$

$$= 1621.14.$$

This measure is much more widely used than the range because it has desirable mathematical properties, some of which will become apparent in this book.

The square of the Greek symbol sigma, σ, has been used to represent the variance even though some readers may not be familiar with Greek letters. The reason is that σ is widely used for this purpose in statistical literature. Other Greek letters have been used in this book for the same reason.

The square root of the variance is frequently taken because it expresses the spread in terms of the original, basic units of measurement. For example, in the taxpayer population, the observations are in terms of dollars, the range $R_y = 140.0$ is in terms of dollars, but $\sigma_y^2 = 1621.14$ is in terms of *squared* dollars. Consequently, the square root $\sigma_y = 40.26$, is in terms of the original unit, dollars, and as a result, is frequently easier to interpret.

Sometimes the quantity,

$$S_y^2 = \frac{1}{9-1} \ [(60-103.56)^2+(72-103.56)^2+ \cdots +(200-103.56)^2]$$

$$= 1823.78,$$

is used in place of σ_y^2. The only difference between S_y^2 and σ_y^2 is the divisor which is equal to the number of units, 9, for σ_y^2 , and one less than this, $9 - 1 = 8$, for S_y^2. The reason for introducing S_y^2, when it is so similar to σ_y^2, is that it "arises naturally" in a number of theoretical calculations and so a separate notation is convenient.

To summarize this section, an illustrative target population of taxpayers was set up. By discussing this population as if it were known completely, we will be able to better understand the results of our actions when, in the later chapters, we draw samples of these taxpayers in various ways. The measurements on each taxpayer are the actual income (y) and the reported income (x). Reported income is presumed known for every taxpayer, but actual income will be known only for a sample of audited taxpayers. Most interest is in the actual income (y). Later in the book, we shall see how to use the known information on x to improve the information on y.

Keeping in mind that only a sample subset of the target population will be observed in practice, some interesting measurable characteristics called *parameters* of the target population were introduced. Two of these, the mean and median, are measures of the location of the middle of the population, and two others, the range and variance, are measures of spread, or variability, in the population.

We shall return to this taxpayer population many times.

Exercises

3.2.1   It turns out that taxpayers T_1, T_2, T_3, T_5, and T_9 are Democrats and the rest are Republicans. This means that the proportion of Democrats in the population is 5/9. Define

$$z_i = 1 \quad \text{if } taxpayer\ i\ is\ a\ Democrat$$
$$= 0 \quad \text{if } taxpayer\ i\ is\ a\ Republican.$$

Then show that

$$\overline{Z} = \frac{1}{N} \sum z_i = 5/9 = P,$$

the proportion of Democrats in this population. The variable z_i effectively "counts" the number of Democrats. For this reason such a variable is sometimes called a "count" variable.

3.2.2   Referring to the previous exercise, show numerically or algebraically that

$$\sum_{i=1}^{N} (z_i - \overline{Z})^2 = \sum_{i=1}^{N} z_i^2 - N\overline{Z}^2 .$$

Hint: Complete the square on the left-hand side of the preceding equation and then sum each of the squared terms.

3.2.3   Referring to the count variable in Exercise 3.2.1 show that

$$\sigma_z^2 = \frac{1}{N} \sum (z_i - \overline{Z})^2 = \frac{20}{81}$$

and that

$$S_z^2 = \frac{1}{N-1} \sum (z_i - \overline{Z})^2 = \frac{5}{18} .$$

3.2.4   Verify numerically that σ_z^2 and S_z^2, respectively, can be alternately calculated by

$$PQ = \frac{20}{81}$$

and

$$\frac{NPQ}{N-1} = \frac{5}{18} .$$

3.2.5   Use the earlier results of exercises to show that

$$\sigma_z^2 = PQ \quad where \quad Q = 1 - P$$

and

$$S_z^2 = \frac{NPQ}{N-1}$$

and z_i is the count variable of Exercise 3.2.1.

3.2.6   Subtract 100 from the actual income of each taxpayer in Table 3.2, that is, $w_i = y_i - 100$ for $i = 1,2,...,9$. Calculate $\overline{W} = (1/N) \sum_{i=1}^{9} w_i$ and show that

$$\overline{W} = \overline{Y} - 100 = 103.56 - 100 = 3.56.$$

(Notice that this same result holds for the addition or subtraction of any constant value to all of the observations. The value of this general result is that it can often be used to simplify calculations.)

3.2.7   Referring to Exercise 3.2.6, calculate σ_w^2 and show that

$$\sigma_w^2 = \sigma_y^2 = 1621.14$$

and

$$S_w^2 = S_y^2 = 1823.78 .$$

(Notice again that this same result holds for the addition or subtraction of any constant value to all of the observations. That is, such an operation does *not* change the variance.)

3.2.8   Multiply each of the actual incomes in Table 3.2 by 10, that is, $w_i = 10y_i$ for $i = 1,2,...,9$. By numerical calculation verify that $\overline{W} = 10\overline{Y}$, $\sigma_w^2 = 10^2\sigma_y^2$, and $\sigma_w = 10\sigma_y$.

3.2.9   If $w_i = ky_i$, $i = 1,2,...,9$, where k is a nonzero constant, formulate two general results based on the findings of Exercise 3.2.8.

3.2.10   If the numerical values of the actual taxpayer incomes are changed by the formula

$$f_i = 50 + \frac{10(y_i - \overline{Y})}{\sigma_y}, \quad i = 1,2,...,N ,$$

show that the f_i values have mean 50 and variance 10. (This is a useful result. For example, the results of two different tests may be scaled in this way to create scores which are more readily comparable because they have the same mean and variance.)

3.2.11   Averages are terribly misleading. It makes no sense to discuss two and *a half* children per family; there are *no* such families! Discuss this statement.

3.3 Smooth Approximations to Target Populations

If a set of data is large, listing all the units and their associated measurements is a cumbersome task. So, it is convenient to summarize large data sets. By a data summarization, we mean an informative, reasonable, short way to describe a set of data, other than by simply listing all the individual observations. An example of data summarization is given in this subsection. In this example, a smooth curve is used to represent a moderately large, discrete set of data. The perceptive reader may realize that most statistical methods are correctly described as data summarizations, and that generally data summarization is a major goal of these methods.

The taxpayer population consists of only nine persons. This population is already small enough that it can be readily understood and studied without further summarization. So for this illustration, we are going to use a larger population.

The data are measurements derived from 500 United States corporations. Each measurement, d, is the absolute value of the difference between the *actual* percentage rate of return, A, earned by the company in 1970, and a statistically derived *forecast*, F, of that rate of return, that is, $d = |F - A|$. The details of the method of forecasting need not concern us here. It is sufficient to know that we have measurements associated with each of 500 United States corporations, and that these measurements are errors in forecasting their rates of return.

The 500 forecasting errors are not listed here. The reason is easy to understand. It is almost impossible to glean any overall insight from a listing of 500 numbers. If the names of the companies were included along with the forecasting errors, such a listing might be helpful in specific cases, but the view we take is that we want to study the population of 500 forecasting errors collectively.

From discussion of the taxpayer population, we know that the mean is one convenient measure of the center of the population. In this forecasting example, it has been calculated that

$$\overline{D} = \frac{1}{500} \sum_{i=1}^{500} d_i = 3.01,$$

where d_i is the error for the *ith* company and capital sigma, $\sum$, denotes the summation of the 500 errors, d_i. So, on the average, the forecasts are off by an absolute amount of 3.01. In contrast, the median forecast error is equal to 1.16, which is considerably less than 3. Now this is interesting because in the last section we learned that the mean is more sensitive to large and small values than the median. Consequently, in this set of forecast errors, it appears that there are a few large errors which are pulling the mean error up, but which are not having the same effect on the median.

We also learned from the taxpayer example, that the variance is a useful measure of the spread of the population. In this forecasting example,

$$\sigma_d^2 = \frac{1}{500} \sum_{i=1}^{500} (d_i - \overline{D})^2 = 6.99.$$

This mean and variance information is certainly helpful, and tells us more about the population than we could learn easily from the listing of 500 numbers. But there is a further, useful summarization that is within our reach. It is called a histogram.

To construct a histogram, group the raw observations together in intervals. In this forecasting example, the intervals are taken to be of length 0.25, that is, a forecasting error of 1/4 of one percent. This means that the errors are grouped in the intervals, {0 to 0.24}, {0.25 to 0.49}, {0.50 to 0.74}, {0.75 to 0.99}, and so on.

Next, the forecast errors falling into each of the intervals are counted. It turns out that there are 36 forecast errors that are between 0 and 0.24 in magnitude. Similarly, there are 77 forecasts that are in error by an amount between 0.25 and 0.49. The counts for all of the interval groupings appear in Figure 3.2a.

The plot shown in Figure 3.2a is called a histogram. It is constructed by drawing a rectangle over each of the class intervals. The base of each rectangle is equal to the width of the interval, and its height is *proportional* to the number of forecasting errors that fall into that particular interval. For example, the rectangle constructed on top of the interval 0 to 0.24 has a height of 36/500 units. The rectangle over the interval {0.25,0.49} has a height of 77/500. The remaining rectangles are constructed similarly. The number of errors in each of the intervals is printed at the top of the corresponding rectangle in Figure 3.2a.

If we adopt the convention that the length of each of the intervals is one unit, then the *area* of each rectangle, (base × height) is simply equal to the height. For example, the area of the rectangle over the interval 0 to 0.24 is 1×36/500 = 36/500. Since there are 500 measurements in all, the sum of the area of all rectangles is 1. This turns out to be a useful convention because we can now informatively add areas. For example, by adding the areas (see Figure 3.2a) to the right of the point 2.0, we can tell that (22+23+12+···+1+1+2)/500 = 126/500, or 25.2%, of the forecast errors are bigger than 2. Or alternately, that 74.8% are less than 2. In Chapter 8 we use areas in this way to develop some particularly helpful insight.

It was stated earlier that data summarization is one of the major goals of statistics. The histogram is one such a summarization. The histogram plotted in Figure 3.2a requires only the 29 class intervals and the counts within each of those intervals. It does not directly require all 500 observations. This is a substantial reduction in the size of the data handling and many times such a reduction is very useful. However, the data summarization can be pushed even further.

It is easy to imagine that, with more forecasting errors and smaller intervals, the histogram of Figure 3.2a might be adequately described by the smooth curve superimposed on it in Figure 3.2b. In this particular case, it can be argued that either the first or second interval is not too well represented by the curve; nevertheless, intuitively, one can still imagine using such a smoothing. If this approach seems reasonable, then all that is needed to describe, or model, the 500 observations, is the *equation* of the smooth curve drawn there. Not even the intervals and counts required by the histogram are necessary, much less the 500 original numbers.

Sometimes, it is very helpful in statistical work to assume that a population can be adequately described by a smooth curve. There are many such curves, with different shapes, that are available for use. The curve drawn in Figure 3.2b is called a *gamma* density. It is an important curve in statistics and has a known formula to describe it. This formula is not given in this book for the simple reason that it is not needed. In Chapter 8 another smooth curve, the normal density, which is even more important than the gamma, is introduced. The formula for the normal density is given in Chapter 8.

There are two important points which remain to be made in this section. The first is that the smooth curve was drawn in (Figure 3.2b)

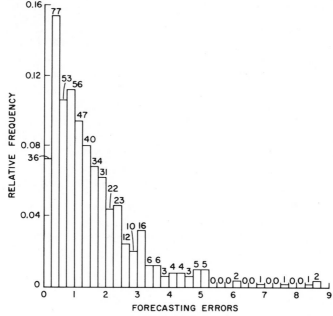

Figure 3.2a. Errors in forecasting corporate rates of return.

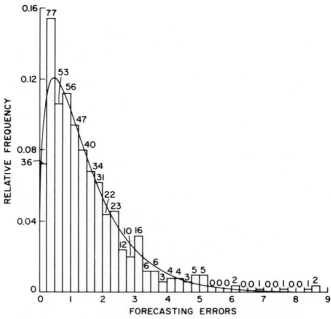

Figure 3.2b. Smooth approximation to forecasting errors.

by reference to the data. We argued that the smooth curve "looked reasonably like" the histogram of the actual data. Sometimes, it is useful to *assume* that the population is adequately described by a smooth curve, *without* any reference to real data. Of course, any subsequent analysis is dependent upon this assumption. If it is not realistic, misleading results may emerge.

The second and final point of this section relates to the informational content of data summarizations. With only the histogram or the equation of the smooth curve, one cannot return to the individual 500 points. Thus some information has been lost. But it is important to realize that some information has also been *gained.* The information gained is the *shape* of the histogram (and the curve).

In the forecasting example, notice that errors of 0.25 to 0.49 in magnitude, are more common than errors of the smaller magnitude, 0 to 0.24. How can this be? Would it not be more reasonable to expect that the most common errors would cluster around and be close to zero? It is clear that this does not happen. The forecasting mechanism seems to have a peculiarity about it. Although we are not going to discuss this forecasting difficulty here, it is important for us to realize that the data summarization made the peculiarity quite evident.

The histogram also shows that the data are stretched to the right. Of course, this was also noted earlier, when we observed that the mean forecast error (3.01) is more than twice the median error (1.16). This skewness reflects the fact that the absolute forecast errors cannot be less than zero, but can be indefinitely large.

In summary, we have seen that representing populations by smooth curves not only may save effort but may also provide new insight. We return to this set of forecasting errors in Chapter 8.

Exercises

3.3.1 The United States Department of Commerce reported in the 1971 Statistical Abstract of the United States that family income for 1969 was distributed as follows.

Income Level	0 $1000	$1000 $1999	$2000 $2999	$3000 $3999	$4000 $4999
Percent	1.6	3.1	4.6	5.3	5.4

Income Level	$5000 $5999	$6000 $6999	$7000 $9999	$10,000 $14,999	over $15,000
Percent	5.9	6.4	21.7	26.7	19.2

Select a set of interval midpoints and use them to calculate the mean family income. (Use $20,000 as the midpoint of the largest grouping.) Also, calculate the variance of family income.

The Department of Commerce gives the median family income as $9433. From the table of incomes we see that 32.3% of the families have incomes below $7000 and that 54.0% have incomes below $10,000. Use this information to estimate the median family income and compare it with the figure published by the Department of Commerce.

Compare your calculated mean family income with the median income of $9433 and explain the difference between them.

3.3.2   In Exercise 1.3.2 a table of supposedly random numbers was given. Make a histogram showing the frequency with which each of the digits 0, 1, 2, 3, 4, 5, 6, 7, 8, 9 appears in that table.

3.3.3   In Exercise 1.3.3, a table of numbers was generated using randomly tossed coins. Make a histogram showing the frequency with which each of the digits, 0, 1, 2, 3, 4, 5, 6, 7, 8, 9, appears in your table.

3.3.4   In the table of numbers of Exercise 1.3.2 make a histogram of the first 50 numbers only. Then repeat the process for the last 50 numbers. Does this suggest to you that somewhere in the transcription of this table a systematic error was introduced? If so what was it?

3.3.5   If the table of digits in Exercise 1.3.2 is "random" any adjacent pair of digits is just as likely to have the bigger of the two digits first as second. Go through the numbers of the table in order and write a "+" every time a bigger number follows a smaller one and "−" for the reverse. So for example, the sequence 78942 would lead to $+ + - -$.

Does the resulting sequence of $+$'s and $-$'s suggest that anything else is peculiar about this supposed table of "random" numbers? What is it?

3.3.6   Exercise 3.3.5 suggests that the random numbers of Exercise 1.3.2 seem to appear in groups in which the digits get larger from left to right. Also, the results of Exercise 3.3.4 seem to suggest that some "0"'s had been mistakenly written as "8"'s. Can you combine these two pieces of knowledge to infer *which* "0"'s were erroneously transcribed as "8"'s? Change the erroneous "8"'s back to "0"'s to obtain a "corrected" table of random numbers.

3.3.7   Construct a histogram of the corrected table of digits obtained in Exercise 3.3.6. Does it seem reasonable that this set of digits was really randomly generated?

3.3.8   There is yet another peculiarity in the corrected table of Exercise 3.3.6. It relates to the fact that the numbers appear in groups of 5 with no repetitions. Can you find the peculiarity? Describe it in detail.

3.3.9   As a result of the exercises in this section, it is now clear that the table contained in Exercise 1.3.2 has certain peculiarities. In the light of these peculiarities review your original answer to Exercise 1.3.2. If your original view of the "random" numbers of Exercise 1.3.2 did not include the peculiarities that we have subsequently discovered, keep this lesson in mind on those occasions when you are assured that a set of data handed to you for analysis is certainly a "good" random sample.

IMPORTANT NEW IDEAS

target population	population units (elements)
frame	population size
distribution	mean
median	range
variance	standard deviation
parameters	histogram
gamma density	normal density

CHAPTER 4

SAMPLES

In the last chapter, the idea of a target population was introduced along with the suggestion that information about a target population would be obtained by observing a sample of it rather than the whole target population. In this chapter, some general characteristics of samples are discussed, although the statistical connections between samples and target populations are not introduced until Chapter 5.

4.1 Samples are Necessary

Samples are used every day to give information about larger groups, and the reason for this frequent use is that there is no alternative. Samples *must* be used. Teachers use examinations to sample the knowledge that students have absorbed because there is no way of measuring everything that a student has learned. Also how can direct impressions of a neighboring state be based on anything other than the occasional (sample) visits made to that state? Since there are endless similar examples of this implicit use of samples, it is clear that samples are both necessary *and* in widespread use.

The widespread use of samples occurs even though misleading conclusions may result if the sample is not similar to the target population. Also, different samples from the same population may be different from each other as well as different from the target population. As we shall see, *statistical samples,* which are samples selected by a *specified random process,* can provide objective measures of this between-sample variability and relate it to the target population.

Statistical samples are necessary for the same reasons that the informal samples described previously are necessary. In fact, whether a sample is statistical or not has little to do with the *necessity* of sampling as opposed to the 100% observation of every population unit, sometimes called a census. An election candidate cannot feasibly assess his chances by interviewing *every* voter in his electoral area. Only a sample is economically reasonable. And it is certainly not sensible for an engineer to determine light bulb longevity by the destructive testing of every bulb coming off a production line.

All in all, it is not very hard to see that in many situations samples *must* be used. We cannot do otherwise, and this is true whether the sample is an informal one or a statistical one. So like it or not, sample we must; although as it turns out, liking it is easy, because there are other very good reasons for using samples rather than 100% observation.

4.2 Samples are Cheaper and Faster

Samples are cheaper and faster than censuses because they are usually only a fraction of the size. Samples are often less than 1% of the size of the target population and are nearly always less than 5%. Occasionally, samples may be as large as 20%, but these do not seem to be frequent and so far seem always to be associated with data stored in computers or very small populations. But whatever the situation, the reduction in the magnitude of the work can be very worthwhile, and it would still be, even if the ratio of sample to census were much larger than it usually is in practice.

Smaller size implies reduced manpower requirements. It is true that the people required to design and process statistical samples correctly are specialized personnel and are, as a result, somewhat expensive. But usually such costs are small when compared to the much larger cost reduction associated with reduced size.

Data processing costs are also less. This is still true today even though cases now arise in which data gathering is electronic and the cost may appear to be about the same whether the data are sampled or not. The fact is that larger sets of data require more attention by trained people, and most analysis requires longer computer running time.

Total project time is smaller for smaller projects. Since some setup time is independent of size, the reduction in time is not always as great as the reduction in the number of observations but it is nevertheless usually significant. Time is money.

4.3 Samples Are More Accurate

Both sampling and nonsampling "errors" are discussed in statistics. Unfortunately, the word "error" can lead to confusion because in general use it suggests a goof, a mistake, but in the jargon of statistics it is not necessarily so.

"Sampling error" means sampling variability, and refers to the fact that the vagaries of chance will nearly always ensure that one sample will differ from another, even if the two samples are drawn from exactly the same target population in exactly the same random way. These potential differences are measured by "sampling error," but a mistake, or error in the goof sense, is certainly not implied. All that is meant is that the estimates differ from each other solely as a result of random selection.

On the other hand, nonsampling errors may indeed involve goofs and blunders. "Nonsampling errors" are those factors that are not strictly random in the sense described in the previous paragraph, and yet which may cause the sample to inaccurately represent the target population. For example, a nonsampling error would be introduced if a zero was recorded whenever an eight was specified. As people experienced in data processing know, there are many many ways in which nonsampling errors can be introduced.

Sampling errors are discussed throughout this book, but nonsampling errors are discussed only in certain sections, such as this one and in Section 6.2. In this section, the single point is that the bigger the study the bigger the chance that nonsampling errors will appear and create all sorts of difficulties.

Evidence exists that the larger the data set is the larger the *percentage* of nonsampling errors will be! Unfortunately, complete discussion of nonsampling errors is both broad and complex and cannot be exhaustively included here. Instead we shall appeal to intuition to argue that small projects are much more controllable than large ones. To do this consider your checkbook and the difficulty you have balancing it; then imagine what could happen doing the same thing on a larger scale, say for a small business, or worse yet for a large business! Perhaps citing the mess that we consistently make of our checkbooks seems overly simplistic in comparison with sophisticated large scale surveys but there is nothing about being "official government" or "company private" which makes data any less susceptible to error, in fact they could well be more susceptible because all too often no one person really cares. A special U.S. Census Bureau study found some nonsampling errors that were 10 times the magnitude of sampling errors, and an important book by Oscar Morgenstern,[1] "On the Accuracy of Economic Observations," has some terrifying examples of data inaccuracies.

1. Morgenstern, O., On the Accuracy of Economic Observations, Second Edition, Princeton Univ. Press, 1963.

Skeptics of sampling principles, who argue for 100% sampling, or even for very large samples, are usually ignorant of the fact that non-sampling errors often create very serious problems. *Samples almost always are more accurate.*

4.4 Statistical Samples and Judgement Samples

A *judgement sample* is one picked from the target population by the subjective decision of an individual. A *statistical sample* is one which is selected by a *specified method* of random selection.

The advantage of a statistical sample is that it permits the objective measurement of sampling variability. This statistical measurement is the topic of this book. Such measurement is impossible with most judgement samples. The only exception occurs in circumstances in which judgement samples are repeatedly used to sample the same kind of population. However, this is not to say that judgement samples are inaccurate, the problem is with the *measurement* of the sample accuracy.

To illustrate, a lumber company needed to estimate the number of board feet left in an area after the area had been lumbered. Up until that time the company had relied upon a skilled employee who developed estimates from the study of aerial photographs. Over years of repetition, the company felt that it had a pretty good idea of how accurate the expert was. The problem was that he was due to retire and no one else seemed ready to assume the task, or at least the company seemed unwilling to trust any new person. A statistical sample seemed to be just the thing. But it was an interesting lesson to discover that the judgement sampler was indeed an expert, and it took considerable effort to develop a statistical sampling plan with comparable accuracy.

What then are the advantages and disadvantages of statistical sampling over judgement sampling? First, the accuracy of judgement samples cannot usually be determined. They are not necessarily inaccurate, but if they are accurate the accuracy is usually unknown and depends upon the expertise of a specific individual. Such experts are rare. The lumbering company had only one. Furthermore, consider the potential difficulties if the expert turned out to be a fraud. It could take years to find out. Statistical samples do not have these problems. Samples can be drawn and their accuracy determined by people who are much more readily available.

4.5 Sample Satisfaction

The proof of the pudding is in the eating, and statistical sampling has been used extensively and successfully in many areas. These areas include epidemiology, industrial quality control, work sampling, accounting, sociology and economics. There are very many examples, the books by Wallis and Roberts,[2] and by Yates,[3] give references to many. However, perhaps nowhere is the use of sampling more impressive than it is in accounting, because in these studies sampling methods are actually used in place of detailed financial settlements, sometimes involving large amounts of money. There are two celebrated examples.

The first example is a plan worked out by the domestic airlines for the allocation of ticket revenues of passengers using more than one airline.[4] In such cases revenue accrues to each of the airlines involved, but the fare is collected only by one. Traditional accounting would call for a settlement on an item-by-item basis. But it is not done this way; instead it is done on a sample basis. The reason is that the cost reductions obtained by processing only a sample of the tickets outweigh the benefits to any of the airlines of processing every ticket. The sample is less than 10% of the total number of involved tickets and all the participants are happy with the scheme.

The second celebrated example is in many ways analogous to the first. It involves telephone calls that are carried over the lines of two or more different telephone companies. In these cases, revenue is due each of the carrier companies but is collected only by one. But as in the airline ticket case, the prospect of keeping records and dividing the revenue for the millions of such calls individually is staggering to contemplate. So the telephone companies also use sampling techniques as a basis for their settlements. There seems to be little reason to suspect that any company would be involved in these plans if it were not in its own interest to do so. These are clear examples of the proof of the pudding.

---

2. Wallis, W. A., and Roberts, H. V., Statistics; A New Approach, Free Press of Glencoe, 1956.
3. Yates, F., Sampling Methods for Censuses and Surveys, Second Edition, Hafner Pub. Co., 1953.
4. Dalleck, Winston, C., Intercompany Account of Settlements by Means of Sampling. Proceeding of the Conference of the Administrative Applications Division, American Society for Quality Control, 1959.

IMPORTANT NEW IDEAS

samples   statistical samples
sampling error   nonsampling error
judgement sample

CHAPTER 5

HOW MISLEADING CAN AN HONEST SAMPLE REALLY BE?

5.1. An IRS Sampling Plan

In practice, *reported* incomes (x) are known for every taxpayer. But it is the *actual* incomes (y) which are of most interest, and these are known for sure only for those taxpayers who are audited. Presumably, there are two reasons for the interest in actual income. The first is that the IRS uses audited incomes to adjust taxes payable. The second reason is that the totals of actual incomes and reported incomes can be compared. Then if the total reported income is very much less than the total actual income, it would pay the IRS to devote more resources to the audit operation. On the other hand, if these totals are close to each other, then the audits would not be of great importance except perhaps as a psychological control on taxpayer honesty. It seems unlikely that total actual income would be substantially less than total reported income.

For reasons of high cost, not every taxpayer can be audited, so information on actual incomes must be obtained on a sample basis. This in turn, implies that estimates of total actual income must also be made from this same sample.

Using the taxpayer population discussed in Chapter 3, suppose that enough resources are available to conduct three audits; how should the three taxpayers be selected? First of all, it seems politically unwise to let the biggest taxpayer escape an audit simply as a result of the vagaries of chance, so let's agree that no matter what else is done, he will be audited. Fortunately, this deliberate selection turns out to be a good statistical move also. This special treatment of the big taxpayer is deferred until Section 11.2, so for now consider the assignment of the other two auditors.

With the largest taxpayer already selected, two other taxpayers must be chosen from the remaining eight. That is, a sample of size $n = 2$ will be drawn from the reduced population of $N = 8$ taxpayers.

50

This new reduced target population is specified in Table 5.1. The only difference between this taxpayer population and the original one is that taxpayer 9 does not appear. This implies that the reduced population has means and variances that are different from the complete population. These new parameters are given at the bottom of Table 5.1.

TABLE 5.1

REDUCED TAXPAYER TARGET POPULATION

$N = 8$

Taxpayer	Actual Income	Reported Income
T_1	$y_1 = 60$	$x_1 = 50$
T_2	$y_2 = 72$	$x_2 = 56$
T_3	$y_3 = 68$	$x_3 = 66$
T_4	$y_4 = 94$	$x_4 = 76$
T_5	$y_5 = 90$	$x_5 = 90$
T_6	$y_6 = 102$	$x_6 = 100$
T_7	$y_7 = 116$	$x_7 = 112$
T_8	$y_8 = 130$	$x_8 = 110$

$$Y = \sum{}^1 y_i = 732.0 \quad X = \sum{}^1 x_i = 660.0$$

$$\bar{Y} = Y/N = 91.50 \quad \bar{X} = X/N = 82.50$$

$$\sigma_y^2 = 515.75 \quad \sigma_x^2 = 507.75$$

$$S_y^2 = 589.43 \quad S_x^2 = 580.29$$

At this point some remarks on notation are necessary. First, the reported incomes, x, are not brought into the analysis until Chapter 12. Consequently, the subscripts x and y are dropped until Chapter 12. In the meanwhile, no confusion is likely to arise when S^2 and σ^2 are used for S_y^2 and σ_y^2. Also until Chapter 12, both Y and $\bar{Y}$ refer to the reduced population of size $N = 8$.

---

1. As in Chapter 3, capital sigma refers to summation; for example, $Y = \sum y_i = y_1 + y_2 + \cdots + y_N$.

First, notice that with the big taxpayer removed, the mean incomes are substantially lower; actual income has fallen from 103.56 to 91.50 and reported income from 92.78 to 82.50. These differences certainly make it clear that the specially treated taxpayer must not be forgotten when estimates are made for the complete $N = 9$ population. How to do this is discussed in Section 11.2.

The variance of the reduced population is also much smaller than the variance of the complete population. For actual income, it has dropped from 1621.14 to 515.75. Now intuitively, this drop in variance seems good because a smaller variance should make the population easier to study. To illustrate with an extreme example, imagine sampling to estimate the average number of cylinders in a standard Volkswagon Rabbit automobile. Since every Rabbit has exactly four cylinders, there is no variance at all in this population. As a result, a single observation will tell all there is to know. Of course, a population with zero variance has almost no practical interest, but at least superficially, it appears that the special handling of the big taxpayer will help the IRS sampling study by this reduction in population variance. We'll see later that this is exactly what happens.

The sampling of two taxpayers from the remaining eight will be done by random selection. To do this easily, place eight discs, one for each taxpayer, number side face down on a table, then mix them thoroughly and select one of the discs. In this way, each taxpayer has the same equal chance of selection. Next, *without* replacing the first disc, select a second one, so that the remaining seven taxpayers have an equal chance of entering the sample. This process gives a sample of size $n = 2$ in such a way that each of the eight taxpayers has an equal chance of selection. Furthermore, each pair of taxpayers has an equal chance of selection. This procedure is called *simple random sampling without replacement*. Other sampling schemes are discussed later in the book.

The actual use of discs is not central to the sampling process described above. Discs are used for descriptive purposes. Any method of random selection, such as the use of random number tables,[2] which gives each taxpayer an equal chance to enter the sample is just as valid.

In the simple sampling scheme above, there are eight taxpayers that could be selected on the first draw and seven on the second. This means that there are $8 \times 7 = 56$ possible selections. But these selections

2. The Rand Corporation "A Million Random Digits with 100,000 Normal Deviates," The Free Press, Glencoe, Illinois, 1955.

are ordered, and as far as we are concerned, the sample $\{T_1, T_7\}$ (say) is the same as the sample $\{T_7, T_1\}$. Consequently, there are $(8 \times 7)/2 = 28$ *different* samples that could arise. Furthermore, each one of these samples is just as likely to arise as any other. The sample $\{T_2, T_6\}$ is just as likely as the sample $\{T_3, T_5\}$.

The number of ways in which two taxpayers can be selected from eight can also be written as $\dfrac{8!}{6!2!} = 28$ which is sometimes also compactly written in the combinatorial notation $\binom{8}{2}$. The notation $n!$ indicates the factorial product $n(n-1)(n-2) \cdots 2 \cdot 1$.

In practice, only one sample will be selected and observed (audited). For illustration, suppose that the sample includes taxpayers 2 and 7, then the audits will find $y_2 = 72$ and $y_7 = 116$. These sample numbers make up a sample distribution, which in this example happens to be trivially simple with only two observations. Nevertheless, just as with the target population, this sample distribution can be plotted and its mean and variance calculated. For this particular sample, the mean is $\bar{y} = (72+116)/2 = 94.00$ and the sample variance is $\hat{\sigma}^2 = \{(72-94)^2+(116-94)^2\}/2 = 484.00$. Different samples will usually have different sample means and different sample variances.

At this point some more remarks on notation are necessary. The small case, $\bar{y}$, is used for the *sample* mean. Similarly, small case $\bar{x}$ is used to denote a *sample* mean. However, a "hat" is used, $\hat{\sigma}^2$, to indicate the *sample* variance. These usages are consistent with general usage in the field of statistics.

Since only the sample mean and variance are known in practice, while the true mean and variance are not, it seems natural to use $\bar{y} = 94.00$ and $\hat{\sigma}^2 = 484.00$ as estimates of the unknown values of $\bar{Y} = 91.50$ and $\sigma^2 = 515.75$. Is this a wise thing to do? For this sample, the sample mean is off by an amount, $\bar{y} - \bar{Y} = 94.00 - 91.50 = 2.50$; how much bigger might this difference be if the process were repeated and a different sample selected? Or how much smaller? These are fair questions even if the second sample is drawn using *exactly* the same method as the first, because the random nature of the selection process will ensure variations in which sample is actually selected. The question is, "How misleading can an honest sample really be?" It is answered in the next section. A discussion of dishonest samples appears in Section 10.3.

Exercises

5.1.1   If there are $N = 11$ city blocks how many different samples of size $n = 3$ can be selected by sampling with equal probability without replacement?

5.1.2   Show that

$$N(N-1)(N-2)...(N-n+1) = \frac{N!}{(N-n)!} .$$

5.1.3   Show that the number of ways of selecting a sample of n city blocks from a total number of N city blocks is

$$\frac{N(N-1)...(N-n+1)}{n!} .$$

5.1.4   If there are 11 persons, how many different ways are there of dividing them into two groups, one of size 3 and the other of size 8?

5.1.5   If there are N persons, how many different ways are there of dividing them into two groups, one of size n and the other of size $N-n$?

5.1.6   Show that $\binom{N}{n} = \binom{N}{N-n}$.

5.2. All the Possible Samples

From the last section we know that there are 28 different samples that can result from a random selection of 2 out of the 8 taxpayers. All 28 possible samples are listed in Table 5.2 along with the associated individual sample means, $\bar{y}$, and sample variances, $\hat{\sigma}^2$. Whenever a sample of size $n = 2$ is drawn from this population, one of these samples *must* result. Consequently, by studying the range of the *possible* values of $\bar{y}$ some idea can be obtained about how far the sample mean could be from the true mean. To illustrate, we see from Table 5.2 that the sample mean could be as large as 123.00 or as small as 64.00, both of which are quite distant from the true mean, $\bar{Y} = 91.50$. If either of these samples is selected, the sample mean, $\bar{y}$, would be as far from the unknown true mean as possible.

The seriousness of being misled by an amount $64.00 - 91.50 = 27.50$ in the case of sample number 2, or $123.00 - 91.50 = 31.50$ in the case of sample number 28, depends largely on the ultimate use of the estimates. But it is certainly reasonable to say that in neither of these two cases have we been misled unfairly because either of these extreme sample means is as likely to occur as any of the more intermediate sample means. However, the chance of being this extremely misled is small because the chance of getting one of these two samples is only 2 out of the total number of 28. Methods for *assessing* how misleading an honest sample can be are discussed in Chapters 7, "How Sure Are You?" and 8, "The Amazing Normal Distribution." In the meanwhile, there are more characteristics of the set of all possible samples that need to be explored.

The set of all 28 equally likely possible values of the mean is formally called *the sampling distribution of the sample mean.* It is plotted in Figure 5.1. It simply describes all the possible values of the sample mean that could result from the sampling scheme being used. The specific form of the sampling distribution depends upon certain factors. For example, the sampling distribution is *derived* from the target population. Consequently, a different target population would result in a different sampling distribution. Furthermore, a different estimate calculated on each sample will also result in a different sampling distribution. For example, in Table 5.2, $\hat{\sigma}^2$ is also computed for each sample; this gives a set of 28 values of $\hat{\sigma}^2$ which is called the sampling distribution of the *sample variance.* Different sample sizes also result in different sampling distributions. This is easy to see, for if $n = 3$ there are $\binom{8}{3} = 56$ different samples that can occur, instead of the $\binom{8}{2} = 28$ when $n = 2$. Finally, as will be discussed later, different methods of sampling also result in different sampling distributions.

The study of a sampling distribution tells how far a sample value may be from the true value. So if the sampling distribution is very much spread out, with a large variance, the sampling procedure is more apt to be misleading than it would be if the sampling distribution has a small variance. It is critical to the understanding of sampling to realize that it is the variance of *the sampling distribution* of an estimator that is important. As discussed in the previous paragraph, the sampling distribution (and hence its variance) is a function of a number of factors, only one of which is the target population. Consequently, the variance of the target population is important only through its indirect effect on the sampling distribution!

TABLE 5.2

SAMPLING DISTRIBUTIONS OF $\bar{y}$, $\hat{\sigma}^2$, and s^2

$N = 8$ taxpayers

All Possible Samples of $n = 2$.

Sample Number	Units In Sample		$\bar{y}$	$\hat{\sigma}^2$	s^2
1	1	2	66.00	36.00	72.00
2	1	3	64.00	16.00	32.00
3	1	4	77.00	289.00	578.00
4	1	5	75.00	225.00	450.00
5	1	6	81.00	441.00	882.00
6	1	7	88.00	784.00	1568.00
7	1	8	95.00	1225.00	2450.00
8	2	3	70.00	4.00	8.00
9	2	4	83.00	121.00	242.00
10	2	5	81.00	81.00	162.00
11	2	6	87.00	225.00	450.00
12	2	7	94.00	484.00	968.00
13	2	8	101.00	841.00	1682.00
14	3	4	81.00	169.00	338.00
15	3	5	79.00	121.00	242.00
16	3	6	85.00	289.00	578.00
17	3	7	92.00	576.00	1152.00
18	3	8	99.00	961.00	1922.00
19	4	5	92.00	4.00	8.00
20	4	6	98.00	16.00	32.00
21	4	7	105.00	121.00	242.00
22	4	8	112.00	324.00	648.00
23	5	6	96.00	36.00	72.00
24	5	7	103.00	169.00	338.00
25	5	8	110.00	400.00	800.00
26	6	7	109.00	49.00	98.00
27	6	8	116.00	196.00	392.00
28	7	8	123.00	49.00	98.00
TOTAL			2562.00	8252.00	16504.00
EXPECTATION			91.50	294.71	589.43
VARIANCE			221.04		

As a result of the features discussed previously, the sampling distribution is the fundamental concept in sampling theory. How to obtain more desirable sampling distributions, estimate their characteristics, and especially reduce their variances, makes up virtually all of this book.

For future use, it is convenient to notice now that *each* of the taxpayers appears in exactly $n/N = 7/28$ of the possible samples. For example, T_1 appears in the first seven samples listed in Table 5.2 and none of the others. In general, T_1 appears in $\binom{N-1}{n-1}$ possible samples. Since there are $\binom{N}{n}$ total possible samples, the proportion of samples in which T_1 appears is $\binom{N-1}{n-1}/\binom{N}{n} = n/N$. The same result holds for each taxpayer.

The ratio n/N is sometimes called the *overall* chance of selection. It gives the proportion of samples in which T_1 (say) appears, either by being selected on the first draw *or* the second. In the sampling scheme used in this chapter, simple random sampling without replacement, each unit in the population has an *equal* chance of entering the sample, 7/28. In some of the sampling schemes which we shall discuss later, the units do not have an equal chance of entry.

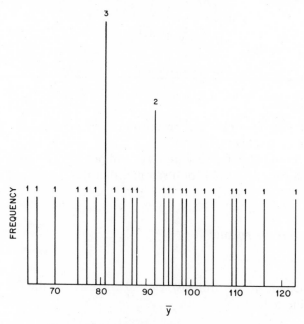

Figure 5.1. Sampling distribution of $\bar{y}$
(two taxpayers from eight).

Exercises

5.2.1   For the small population

$$y_1 = 6, \; y_2 = 2, \; y_3 = 5, \; y_4 = 12, \; y_5 = 10,$$

systematically list all of the possible samples of size $n = 3$ that can be selected by drawing 3 units from 5 with equal probability without replacement.

5.2.2   For the population in Exercise 5.2.1, compute the sample mean for each of the possible samples and plot the sampling distribution of the sample mean.

5.2.3   In the exercises of Section 3.2, it was pointed out that taxpayers T_1, T_2, T_3, T_5, and T_9 are Democrats. Then we showed that

i. $$\bar{Z} = \frac{1}{N} \sum_{i=1}^{N} Z_i = P = \frac{5}{9},$$

ii. $$\sigma_z^2 = PQ = \frac{20}{81},$$

iii. $$S_z^2 = \frac{NPQ}{N-1} = \frac{5}{18}, \text{ where}$$

$$z_i = 1 \quad \text{if } taxpayer\ i\ is\ a\ Democrat$$

$$= 0 \quad \text{if } taxpayer\ i\ is\ a\ Republican$$

and $P = 1 - Q$ is the proportion of Democrats in the population of taxpayers.

Using small case letters for sample values show that analogous formulas hold for a sample, that is,

i. $$\bar{z} = \frac{1}{n} \sum_{i=1}^{n} z_i = p,$$

ii. $$\hat{\sigma}_z = pq, \text{ and}$$

iii. $$s_z^2 = \frac{npq}{n-1}.$$

5.2.4   Calculate p, pq, and $npq/(n-1)$ for each of the 28 samples in Table 5.2.

5.3. Unbiasedness

Since the set of 28 possible means is itself a distribution, a mean and variance can be calculated just as we have already done with both populations and samples.

The average of the 28 sample means is given in Table 5.2. This average value is the value of $\bar{y}$ which we might "expect" to get over repeated trials of the same sampling procedure. As a result, the average value of $\bar{y}$ over all the possible values is sometimes called the *expectation* of $\bar{y}$ and is written, $E(\bar{y})$. It is important to remember that the special notation "E" means that an average is being taken over all the possible values in a sampling distribution.

In the example, $E(\bar{y})$ is 91.5, which is exactly equal to the mean of the target population. This is no accident; it is an important general result which can be proven mathematically. The theorem says that this sampling procedure will yield a sample mean which, when averaged over all the possible values that can occur, is exactly equal to $\bar{Y}$. The characteristic that $E(\bar{y}) = \bar{Y}$ is called the property of *unbiasedness,* and $\bar{y}$ is called an *unbiased estimator*[3] of $\bar{Y}$.

The sample variance $\hat{\sigma}^2$ is *not* an unbiased estimator of the population variance σ^2. The mean of the 28 values of $\hat{\sigma}^2$ is equal to 294.71 (Table 5.2), which is not equal to the true value of $\sigma^2 = 515.75$. So $\hat{\sigma}^2$ is clearly not an unbiased estimator of σ^2. The reason is the divisor of n in $\hat{\sigma}^2$. In Table 5.2 the sampling distribution of s^2 is also given, where s^2 is calculated with a divisor of $n-1$. For example, in sample number 1, s^2 is calculated as

$$s^2 = \frac{1}{2-1}\{(60-66)^2+(72-66)^2\} = 72.00. \tag{5.1}$$

Now observe that the average value of the 28 values of s^2 is exactly equal to 589.43, which is the value of S^2 in the target population, that is, $E(s^2) = S^2$. So s^2 is an unbiased estimator of S^2. This result can also be proven mathematically and is one reason why divisors of $(n-1)$ and $(N-1)$, rather than n and N, are frequently used for the sums of squares of deviations of measurements from their mean.

---

3. When reference is made to the *entire set* of possible sample means, $\bar{y}$ is referred to as an estimator. A particular value of $\bar{y}$ is called an estimate. Similar terminology is used for other estimators, such as the sample median or the sample standard deviation.

Exercises

5.3.1   Calculate $E(\bar{y})$ for the sampling distribution in Exercise 5.2.2 by direct use of the sampling distribution.

5.3.2   The unbiasedness of $\bar{y}$ and s^2 does not depend upon the particular measurements associated with each of the taxpayers. So, for example, the sample mean is unbiased whether the measurement associated with each taxpayer is income or the zero-one variable discussed in Exercise 5.2.3.

Verify numerically that

$$E(p) = P \quad \text{and}$$

$$E[npq/(n-1)] = NPQ/(N-1)$$

by using the results obtained in Exercise 5.2.4.

5.3.3   A single die is tossed. If an even number appears face up, you win $1.00 and if an odd number appears you lose $1.00. Write down all the possible results and find your expected winning.

5.3.4   A single die is tossed. If an even number appears face up, you win $1.00, otherwise you lose 50 cents. Write down all the possible results and find your expected winning.

5.3.5   A single die is tossed. If a "1" or "2" appears face-up you win $1.00, and if a "3," "4," "5," "6" appears you lose 75 cents. Do you expect to win or lose at this game. How much?

5.4. The Variance of the Sample Means

Recall that in Section 5.2, we decided that how misleading a sample mean $\bar{y}$ may actually be is critically dependent upon the dispersion of the sampling distribution. In fact, the variance of the sampling distribution is more important than the variance of the target population. As a result, the contents of this section are extremely important.

Since the sampling distribution of the mean is just a set of 28 numbers (all the possible sample means), its variance can be calculated directly in the usual way by summing the squares of the deviations and dividing by 28. From Section 5.3, we know that this calculation is called the expectation of the squared deviations, $E(\bar{y}-\bar{Y})^2$, but for mnemonic purposes we shall refer to it as $Var(\bar{y})$. Specifically,

$$E(\bar{y} - \bar{Y})^2 = \frac{1}{28}\ [(66.00 - 91.50)^2 + \cdots + (123.00 - 91.50)^2]$$

$$= 221.04 = Var(\bar{y}). \tag{5.2}$$

The direct calculation of $Var(\bar{y})$ can be used to illustrate one of the amazing theorems of mathematical statistics. This theorem says that $Var(\bar{y}) = E(\bar{y} - \bar{Y})^2$ can alternately be computed exactly by the *formula*,

$$Var(\bar{y}) = E(\bar{y} - \bar{Y})^2 = (1 - \frac{n}{N})\ \frac{S^2}{n}. \tag{5.3}$$

Indeed, we can verify that,

$$Var(\bar{y}) = E(\bar{y} - \bar{Y})^2 = (1 - \frac{2}{8})\ \frac{589.43}{2} = 221.04, \tag{5.4}$$

which is the same value as was obtained by the direct method. This extremely useful formula puts the sampling variance, $Var(\bar{y})$, in terms of the known values n and N, and S^2, the dispersion of the *target population*. This means that the calculation of $Var(\bar{y})$ has been shifted from requiring a knowledge of the sampling distribution (Table 5.2) to requiring knowledge of the target population! So, if S^2 were known, the variance of the 28 sample means could be calculated without having to obtain any of the sample means much less all 28 of them. This amazing result is used over and over again in statistical analysis.

In the formula for $Var(\bar{y})$, the multiplier $(1 - n/N)$ is called *the finite population correction* factor, or sometimes simply the fpc. If the population size, N, is very large relative to the sample size, n, the fpc is essentially equal to one and the formula for $Var(\bar{y})$ reduces to S^2/n. This is a form that may be somewhat familiar to readers who have some statistics background. On the other hand, if the sample size is equal to the size of the target population, that is, $n = N$, then $Var(\bar{y}) = 0$. In this case, a zero variance is a reasonable consequence because all units of the target population will have been observed and the sample mean will be exactly equal to the population mean, that is, $\bar{y} = \bar{Y}$.

The general formula for $Var(\bar{y})$ Equation (5.3), shows again, see Section 5.2, that the sampling distribution of $\bar{y}$ is a function of the sample size, the target population, the specific form estimator and the method of sampling. The dependency of $Var(\bar{y})$ on the sample size n, and S^2 (the target population) is shown explicitly in the formula. Also, a different estimator will have a different sampling distribution. For example, 28 sample medians will in general be different from 28 sample means, and so will have a different sampling variance. Finally, the dependency upon the *method* of sampling can be seen by observing that if repetitions were allowed in the sampling, it would be possible to get the sample ($y_1 = 60$, $y_1 = 60$ with $\bar{y} = 60$). Such samples cannot occur in sampling without replacement. Sampling with replacement is discussed in Chapter 11.

The rest of this book is devoted to the understanding and manipulation of these factors that influence the sampling distribution. The objective is the achievement of desirable statistical sampling plans.

Exercises

5.4.1   Calculate $Var(\bar{y})$ for the sampling distribution in Exercise 5.2.2 by direct use of the sampling distribution.

5.4.2   For the population of Exercise 5.2.1 calculate $Var(\bar{y})$ by use of the formula given in Section 5.4. Compare the result with that obtained in Exercise 5.2.2.

5.4.3   What is the relationship between the target population and the sampling distribution of the mean when $n = 1$?

5.4.4   Calculate the range R of the population in Exercise 5.2.1. Then calculate the sample range r for each of the 10 possible samples. Calculate the expected value of r, $E(r)$, and the sampling variance of r, $Var(r)$. Is r an unbiased estimator of R?

5.4.5   In the population of Exercise 5.2.1, calculate the median, Y_{med}. Then calculate the sample median, $\hat{y}_{med}$, for each of the 10 samples in Exercise 5.2.1. Next calculate the expected value of $\hat{y}_{med}$, and its variance $Var(\hat{y}_{med})$. Is $\hat{y}_{med}$ an unbiased estimator of Y_{med}?

5.4.6   For each of the 10 samples in Exercise 5.2.1, calculate $s^2 = \sum(y_i - \bar{y})^2/(n-1)$.

5.4.7   Referring to Exercise 5.4.6 show that $E(s^2) = S^2$.

5.4.8   (a) Using the 10 values of s^2 in Exercise 5.4.6 calculate the 10 values of

$$(1 - \frac{2}{5}) \frac{s^2}{2}.$$

(b) Find the average of the 10 values in (a) and show that this average is equal to the value of

$$(1 - \frac{2}{5}) \frac{S^2}{2} .$$

Can you formulate a more general result from this problem?

5.4.9   Calculate $Var(p)$ directly from the 28 values of p obtained in Exercise 5.2.4.

5.4.10   We know from Section 5.3 that

$$Var(\bar{y}) = (1 - \frac{n}{N}) \frac{S^2}{n} .$$

The results of Exercise 5.2.3 suggest by substitution that

$$Var(p) = (1 - \frac{n}{N}) \frac{1}{N} \frac{NPQ}{N-1} .$$

Calculate $Var(p)$ using the above formula and compare the result with that obtained in Exercise 5.4.9.

5.4.11   The following sample was drawn from a population of size $N = 20$:

$$2, 3, 6, 7, 9 .$$

(a) From this sample can an unbiased estimate of the variance of the mean of a sample of size 7 be made? If not, state why not. If yes, describe how the estimate is made.

(b) Answer the same question as part (a) for a sample of size 3.

5.5. Of What Use Is a Single Sample?

We have seen that there is a great deal to be learned by examination of the distribution of all possible sample means. Now, it is legitimate to ask of what use is a single sample? After all in practice, only 1 of the 28 possible samples will be drawn and that is all the information there will be. Fortunately, as we shall see, there is a great deal that we can learn, even from a single sample.

First of all, with a single sample the sample mean $\bar{y}$ can be calculated and used as a proxy for the unknown true mean $\bar{Y}$. Furthermore, this can be done with some confidence because we know that $\bar{y}$ is unbiased, so that over all possible repetitions of the same sampling scheme $\bar{y}$ will be equal to $\bar{Y}$.

Furthermore, it is possible to estimate the variance of the sample mean even though only one sample is available. To do this it is only necessary to combine two facts that we have already discussed.

The first fact is the mathematical result (Equation 5.3) which puts $Var(\bar{y})$ in terms n, N, and S^2. It is repeated as Equation 5.5.

$$Var(\bar{y}) = (1 - \frac{n}{N}) \frac{S^2}{n}. \qquad (5.5)$$

Now n and N are known ($n = 2$ and $N = 8$ in the taxpayer example), so consideration of $Var(\bar{y})$ comes down to consideration of S^2. But S^2 cannot be known from a single sample because its calculation requires knowledge of the entire target population. But now we can introduce a second useful fact. It is that S^2 can be estimated unbiasedly by s^2; see Section 5.2. This suggests substituting s^2 for S^2 in the formula for $Var(\bar{y})$ to give

$$var(\bar{y}) = (1 - \frac{n}{N}) \frac{s^2}{n} \qquad (5.6)$$

as an estimate of $Var(\bar{y})$. The important point is that $var(\bar{y})$ can be calculated *entirely from the sample*. Notice that, just as we have done in a number of other places, a small case letter has been used to indicate a sample estimate. That is, $var(\bar{y})$ is an estimate of $Var(\bar{y})$.

To illustrate that $var(\bar{y})$ can be calculated from a single sample, if the sample ($y_2 = 72$, $y_7 = 116$) with $\bar{y} = 94.00$ and $s^2 = 968.00$ is drawn (see Table 5.2), then $\bar{y} = 94.00$ is an estimate of the true mean $\bar{Y}(= 91.50)$ and

$$var(\bar{y}) = (1 - \frac{n}{N}) \frac{s^2}{n} = (1 - \frac{2}{8}) \frac{968.00}{2} = 369.75$$

is an estimate of the true sampling variance,

$$Var(\bar{y}) = (1 - \frac{n}{N}) \frac{S^2}{n} = (1 - \frac{2}{8}) \frac{589.43}{2} = 221.04.$$

It is easy to verify that overall possible samples $var(\bar{y})$ is an unbiased estimator of $Var(\bar{y})$. This unbiasedness follows directly from the fact that s^2 is unbiased for S^2, and the fact that the factor $(1 - \frac{n}{N}) \frac{1}{n} = \frac{3}{8}$, is the same for *all* samples. Hence, when s^2 is averaged over all possible samples (Table 5.2), the factor 3/8 simply appears as the same factor multiplying S^2. So referring explicitly to Table 5.2, we see that,

$$E\{var(\bar{y})\} = \frac{3}{8}\,[72+32+\cdots+98]/28$$

$$= \frac{3}{8}\,589.43 = 221.04 = Var(\bar{y}). \qquad (5.7)$$

So as the pieces are assembled, it becomes clear that there is indeed a great deal to be learned from a single sample.

Exercises

5.5.1   (a) Calculate

$$var(p) = (1 - \frac{n}{N})\,\frac{1}{n}\,\frac{npq}{n-1} = (1 - \frac{n}{N})\,\frac{pq}{n-1}$$

for each of the 28 samples of Exercise 5.2.4.
(b) Use the results in (a) to calculate $E(var(p))$ directly and thereby confirm numerically that

$$E(var(p)) = (1 - \frac{n}{N})\,\frac{NPQ}{N-1}\,.$$

5.5.2   A sample of seven residential telephone customers was observed to have bills in the following amounts: $8.00, $9.15, $14.21, $29.53, $8.80, $9.51, $10.22.
(a) Estimate the amount of the average residential bill.
(b) Estimate the variance among bills.
(c) If there are 10,000 customers in the area estimate the total revenue.
(d) Estimate the variance of the estimate in part (c).

5.6. The Terminology of Populations and Distributions

Statistical terminology for populations and distributions can seem complicated. Some *terminologies* sound similar enough that they might be thought to refer so the same thing. Yet, they are not quite the same, and this leaves doubts in the mind of the newcomer. Consequently, a brief review is worthwhile.

The target population is the group about which information is sought. Examples are the entire United States adult population and the population of taxpayers. In some statistical writing, the target population is referred to as simply the population, or possibly the target distribution, or even the parent distribution. However, to keep the confusion to a minimum, neither "target distribution" nor "parent distribution" have been used in this book.

The "*sampling* distribution of the mean" refers to the set of all possible sample means that could be obtained by *sampling* from a target population in a specified manner. A different sample estimate generally has a different sampling distribution. Also different sampling processes and different sample sizes usually result in different sampling distributions. Sampling distributions are *not* referred to as sampling "populations."

Next comes the *sample* distribution. This simply refers to the set of measurements obtained by observation of a particular sample. Each sample generally has a different distribution, so which sample distribution is studied depends upon which sample is actually randomly selected.

Finally, a word about "universes" and "populations." These words are usually used interchangeably, but some authors use "universe" in special ways. For example, "universe" may refer to the *set of units* in the target population and "population" to the set of *measurements* associated with those units. For example, the target population might be the *universe* of New York City residents, while the *population* of heights and the *population* of weights might be two specific measurements of interest. In some ways, this is a worthwhile distinction, but it is not necessary in this introductory book and is therefore not used.

Later in this book, the *sampled* population will also be discussed in contrast to the *target* population. Briefly, the target population is the one that we intended to sample, the sampled population is the one we actually did sample. They are not always the same.

IMPORTANT NEW IDEAS

estimator

expectation

finite population correction

estimate

unbiased estimator

sampled population

simple random sampling without replacement

sampling distribution of the sample mean

sampling distribution of the sample variance

CHAPTER 6

DISTORTIONS OF AN
HONEST SAMPLE

6.1 Bias

The word *bias* is commonly used with vague and controversial meanings, but the statistical meaning of bias is quite precise. Furthermore, a statistical estimator which is biased is not necessarily all bad; it may have some properties that are better than an alternative, unbiased, estimator (see Chapter 12). Nevertheless, the common usage of the word bias seems to work against usage of biased estimators. Since biased estimators are sometimes very good it is worthwhile trying to clarify the topic of statistical biases.

We pointed out earlier (5.3) that an estimator that averages over all the possible sample values to the parameter being estimated is said to be unbiased. In the IRS example, the sample mean $\bar{y}$ is an unbiased estimator of $\bar{Y}$, s^2 is unbiased for S^2, but $\hat{\sigma}^2$ is *not* an unbiased estimator of σ^2. How then does an estimator become biased? What are the different types of bias? And what are the properties of these different types?

Sources of bias can be put into three categories. First, *technical biases* are a result of the specific algebraic form of the estimator. While this type of bias is not discussed until Chapter 12, we have already seen an example of it, because $\hat{\sigma}^2$ is a technically biased estimator of σ^2.

Selection bias is a second category. This type arises, when for whatever reason, a sample is *not* drawn according to the prearranged specifications. The result is that population units are actually selected into the sample with chances other than those specified in the sampling design. For example, the sampling design may specify the equal chance selection of households, but in the actual selection, households with children may be more apt to be enter the sample than households without children. The Case of the Disappearing Children (1.7) is a good example. If adjustments are not made for this kind of distortion, any estimates based on the sample are likely to be biased. Selection biases are discussed in Section 10.3, "Stacking the Deck."

In this chapter, the third source of bias, *measurement error,* is discussed. The order of presentation of these biases is not intended to reflect their relative importance because with any category of bias the possibilities for disaster are great.

6.2 Crude (But Effective) Distortion

The topic of measurement error was actually introduced in Section 4.3. There it was argued that nonsampling errors can dominate sampling errors in large sets of data. It was also stated there, and it is worth repeating, that measurement difficulties are not adequately discussed in this book. Measurement errors are very difficult to discuss in a general way and there are many different, meaningful, special cases. Conceptually, measurement error is easy to understand, but it is difficult in practice because errors can be introduced in subtle ways. Nevertheless, the topic is extremely important so that some discussion of the measurement apparatus, fuzzy definitions, data handling, and possible assistance offered by formal statistics is necessary.

A measuring device can malfunction. The device can be mechanical or human, and the errors generated completely random, mechanically systematic, or even maliciously deliberate. No possibility can be completely discounted. Misunderstood instructions are an excellent source of error. In a study of business offices in which groups of employees were performing essentially the same tasks, observations were made on the number of times each day, f_{ij}, that the ith employee carried out the jth task. The number of hours, t_i, that the employee worked that day was also observed. This meant that observations of the form, $(t_i, f_{i1}, \ldots, f_{ij}, \ldots, f_{ik})$, were gathered for each employee for each day. Unfortunately, repeated initial attempts to fit statistical models to these data failed because the computer refused to perform a division by $\sum_{j=1}^{k} f_{ij}^2$ which it claimed was equal to zero. Mathematically, the only way that Σf_{ij}^2 can be zero is for *each* one of $f_{i1}, f_{i2}, \ldots, f_{ik}$ to equal zero. How could this be? A printout of the raw data displayed observations of the type $(+, 0, 0, 0, \ldots, 0)$ and $(0, +, +, \ldots, +)$, where the $+$'s indicate positive numbers. The interpretation for the first type of number is that some persons were on the job but did no work, which is odd but entirely possible! The second type of observation is even more fascinating because it reveals that some people did work without actually being there!

The trouble revolved around absent employees and their replacements. The erroneous data-gathering logic went as follows. Since that desk position really belongs to "Bill" (the absent employee), the work done at that position ought to be credited to him in spite of the fact that "Anne" (the replacement) took over for that day. The erroneous logic continued by not crediting the time to Bill, after all he was absent; the time should clearly be credited to Anne because she was there! Consequently, observations of the two different types specified in the previous paragraph arrived for analysis. Actually, the situation was even worse than described so far because occasionally employees were absent for only a *half* day. This meant that much data was in error that could not be identified by the presence of zeros. In this case, the human measuring device went badly astray.

A mechanical apparatus is not necessarily any better. In a rainfall study, plots of the data suggested an unbelievably periodic rainfall; see Figure 6.1. And, indeed it was unbelievable, because close scrutiny revealed a systematic malfunction in the measuring device. The rain flow was impeded in such a way that a small container would slowly fill, while keeping the apparatus from measuring any rainfall, and then would spill out, registering a very heavy rainfall in a very short time span. Again the measuring device was introducing measurement biases, but in this case the measuring device was mechanical.

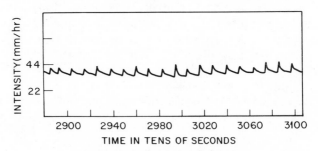

Figure 6.1. Systematic rainfall data
(Gabbe and Freeny).

Interviews with people fall into a special class; many books and papers have been written on the subject of accurate responses. There are many problems. Long and repeated interviews can degrade the answers and induce nonresponse. Also people do many strange things, including forgetting their age and trying to outguess you. Bushel baskets and rulers don't usually have such failings, but it is a mistake to ever relax; after all, the bushel basket could have a hole in it! However, in general, while mechanical measuring devices can indeed fail and act in complicated ways, there seems to be a better chance of accuracy if humans are eliminated from the process.

Imprecise definitions also cause measurement problems. The measurement of unemployment may appear to be simple, but the U.S. Census Bureau has spent much time and effort developing careful, consistent, definitions of the various classes of unemployment. Other countries have not been as careful, with the result that international comparisons are often of doubtful meaning. In other studies, it may not even be apparent that a definitional problem exists. For example, in a study of the demand for telephones in local areas, the time series history of the number of telephones served by a local wire center was under study. The hidden difficulty is that such data are not always comparable through time. Whenever a new wire center is built, a block of customers is "cutover" from the old wire center to the new one. As a result, a given wire center serves a varying geographical area and the historical time series cannot be analyzed directly without some kind of an adjustment for these area "cutovers."

Another major potential source of measurement bias occurs in the handling of the data. Unfortunately, the ease with which such errors are made has nothing to do with the magnitude of their effect. It is very easy to slip a decimal point and wind up analyzing 9049.52 instead of 904.952. If many such errors were made, they would probably be detected, but what if only *one* number is affected? Would it still be found? It is certainly possible to detect such an error, but the degree of insight required is substantially greater than the case in which *all* observations are multiplied by 10. Furthermore, the effect of a single mispunched card is not necessarily trivial, it can for example have a major effect on correlation coefficients, see Chapter 12.

What can statistical methods do for the measurement problem? Perhaps the best suggestion is to *sub*sample the data and remeasure this subset to verify the accuracy of the measurements. Of course, such a scheme would be of no use whatsoever if the original errors were simply repeated. New measurement standards must be applied.

Another statistical approach is to assume that errors are made randomly, so that instead of measuring x, $y = x + e$ is measured, where e is a random error *which averages to zero*. There is a fairly large number of papers in the statistical and economic literature in which this simple assumption is made. Unfortunately, it does not seem to be reasonable in many situations to assume that the errors average zero. Realistically, this type of assumption seems to be a sophisticated guessing game, *unless* some way can be found to objectively develop the relationship between y and x.

Finally, there is an ingenious technique, first proposed by S. L. Warner,[1] that promises to improve measurement accuracy in the sampling of human populations. Since it is well known that people do not always answer questions correctly, the trick of this technique is to get the person to answer a question which is unknown to the interviewer, or for that matter to anyone except the respondent. The objective is to entice the respondent to be more truthful. The procedure is described in the next section.

Exercises

6.2.1   In your own words describe both sampling error and measurement error.

6.2.2   Imagine that you are charged with the task of conducting a school census in your local community. The information is to be gathered by a house-to-house canvas, transferred to punch cards, and turned over for analysis to a local company that has volunteered the use of its computer. Make a list of all the possible sources of inaccuracies, apart from sampling error, that you think might affect any final tabulations. Using this list next make a second list of all the checks for accuracy that you think should be routinely applied in this project.

1. Warner, S. L., Randomized response: a survey technique for eliminating evasive answer bias. Journal of the American Statistical Association, Vol. 60, 63-69, 1965.

6.3 The Sampler's Approach

Suppose that the survey goal is the estimation of the proportion of people who regularly engage in extramarital sex! How is such a question to be asked with any hope of getting an even remotely accurate answer? To ask the question directly and simply hope that the answers are accurate, does not seem reasonable. The question is much too personal. This difficulty leads to Warner's proposal.

Perhaps the accuracy of the answers could be improved if the respondent were assured that no one knew that such a controversial question was being answered!? This thought is the basis of Warner's clever procedure. To show how it works, suppose that question A is "Do you regularly engage in extramarital sex?" and question B is "Is the coin you just tossed a head?" Question A is the question of real interest, question B is a red herring which presumably no one would mind answering. However, it is important that the correct answer to the "irrelevant" question is not known to the interviewer. For example, the question "Is your social security number odd?" should *not* be used for this purpose. The correct answer could very well be known by the interviewer. In some cases such information will permit a clever interviewer to deduce the answer to the sensitive question. We shall describe a simple version of Warner's procedure.

First, the respondent tosses a coin being careful not to let the interviewer see the result. Then, the respondent (not the interviewer) picks a random integer between 1 and 10 by some mechanical means. The person is then instructed that if the resultant number is 1, 2, 3, or 4, he should answer question A, and if the number is 5, 6, 7, 8, 9, 10, he should answer question B. Only the respondent knows the selected random number and hence only the respondent knows which question is being answered. The interviewer knows only the yes or no answer and does not know which question was answered. So, for example, he cannot be subpoenaed and required to testify as to the respondent's sex life.

How can such a procedure yield relevant statistical estimates? Not surprisingly, the method is based on a statistical formula which is written below in a nonmathematical form.

$$
\begin{Bmatrix} Proportion \\ of\ all\ yes \\ answers \end{Bmatrix} = \begin{Bmatrix} Proportion \\ of\ yes's \\ to\ question\ A \end{Bmatrix} \times \begin{Bmatrix} Proportion\ of \\ times\ question \\ A\ is\ asked \end{Bmatrix}
$$

$$
+ \begin{Bmatrix} Proportion \\ of\ yes's \\ to\ question\ B \end{Bmatrix} \times \begin{Bmatrix} Proportion\ of \\ times\ question \\ B\ is\ asked \end{Bmatrix}. \qquad (6.1)
$$

In this formula, the information that *really* is wanted is the proportion of yes answers to question A, the sex question. Fortunately, it turns out that when a large number of respondents have been put through the procedure, everything else in Expression 6.1 is *known,* at least approximately, and consequently, the formula can be solved to find the desired information. To be specific, suppose 1/3 of all answers given to the interviewer are yes's. Now by the random selection of the questions, question A has been answered about 4/10ths of the time and question B 6/10ths; furthermore, the number of yes responses to question B, the coin toss, will be about 1/2. If these known values are substituted in Expression 6.1, it becomes,

$$\frac{1}{3} = \left\{ \begin{array}{c} Proportion \\ of\ yes's \\ to\ question\ A \end{array} \right\} \times \frac{4}{10} + \frac{1}{2} \times \frac{6}{10} \ . \qquad (6.2)$$

which can be solved to give,

$$\left\{ \begin{array}{c} Proportion \\ of\ yes's \\ to\ question\ A \end{array} \right\} = \frac{1}{12} \ .$$

So an estimate of the proportion of persons engaging in extramarital sex can indeed be obtained, while at the same time the privacy of the individual is preserved. A very neat trick indeed.

But does the procedure really work? The hope is that this method will make people more comfortable and hence more apt to respond honestly. But then perhaps it doesn't work that way at all. To give a counterargument, a respondent may realize that some accurate use of the response is going to be made, otherwise the question would not be asked. An unfavorable reaction to this realization could very well defeat the whole procedure. Nevertheless, this technique has been included in this book as one example of the statistical efforts to make measurements more accurate. In practice, a number of ingenious modifications to Warner's basic idea have been proposed, but all have the common feature of a mechanical randomization that only the respondent observes.

Finally, if this chapter has managed to convey a feeling of the enormity of the measurement problem, it has achieved its purpose.

IMPORTANT NEW IDEAS

bias   technical bias
selection bias   measurement bias
random response

CHAPTER 7

HOW SURE ARE YOU?

In the last chapter, we discussed distortions resulting from measurement bias. Unfortunately, these biases usually cause the sample mean, $\bar{y}$, to be farther away from the true mean, $\bar{Y}$, than it would otherwise be. But how far might $\bar{y}$ be from $\bar{Y}$ even if there is no measurement bias? That is, how far can the random variation in the sampling process itself take $\bar{y}$ from $\bar{Y}$? The answer to this question is presented in this chapter.

7.1 Where Are the Most Misleading Samples?

If a sample is drawn by simple random sampling without replacement, as in Chapter 5, how badly might we be misled by using the sample mean, $\bar{y}$, to represent the true mean, $\bar{Y}$? Examination of the sampling distribution of $\bar{y}$ in the taxpayer example, Table 5.2, shows that if sample number 28 (taxpayers 7 and 8) is selected, the biggest overestimate of the true mean occurs. This overestimate is equal to $\bar{y} - \bar{Y} = 123.0 - 91.5 = 31.5$. In Figure 5.1, this sample mean appears at the far right end of the distribution, which is also called the *upper tail* of the distribution. The chance of getting this overestimate is exactly 1 sample in 28, or 1/28.

Sample number 27 (taxpayers 6 and 8) with a sample mean of 116.0 is the second largest overestimate of the true mean. It occurs with the same chance, 1/28, as each of the other 28 possible sample means. Consequently, the chance of getting a sample mean *at least* as far out as 116.0 is $1/28 + 1/28 = 2/28$ because *both* samples 27 *and* 28 are out that far in the upper tail.

Similar statements can be made about the *lower tail* of the distribution. Sample number 2 has a sample mean of 64.0 and hence would miss the true mean by $\bar{y} - \bar{Y} = 64.0 - 91.5 = -27.5$. Sample number 1, with a mean of 66.0, would give the second largest underestimate, by the amount $66.0 - 91.5 = -25.5$. Each of these samples occurs with a *chance,* or *probability* of 1/28, so that the probability of getting a sample with a mean equal to *or less* than 66 is $1/28 + 1/28 = 2/28$.

76

Many similar statements can be made. For example, it can be easily verified that the chance of getting a sample mean equal to or greater than 110.0 is 4/28, and that the chance of getting a sample mean equal to or less than 70.0 is 3/28. Also, statements that refer to the upper and lower tails of the distribution separately, can be combined. For example, the chance of getting a sample mean that is *either* equal to or bigger than 116.0 *or* equal to or smaller than 66.0 is 4/28. The reason is that each of the four samples numbered 1, 2, 27, 28 with means 66.0, 64.0, 116.0, 123.0, meets the stated requirements, and furthermore, these are the only samples that do.

Exercises

.7.1.1   Count the number of sample means in Table 7.1 that are larger than 104.0. Use this count to obtain the probability that in the simple random sampling of two taxpayers from eight that the sample mean actually obtained will overestimate the true mean by at least 12.5.

7.1.2   Count the number of sample means in Table 7.1 that are smaller than 78.0. Use this count to obtain the probability that the sample mean actually obtained will underestimate the true mean by at least 13.5.

7.1.3   Use the results of Exercises 7.1.1 and 7.1.2 to find the probability that the particular sample mean obtained by simple random sampling will *either* be an overestimate of at least 12.5 *or* an underestimate of at least 13.5.

7.1.4   Systematically write down all 56 possible samples of three taxpayers that may be selected by sampling from the population of eight taxpayers with equal probability without replacement.

7.1.5   Calculate the sample median, $\hat{y}_{med}$, for each of the 56 samples listed in Exercise 7.1.4.

7.1.6   Use the results of Exercise 7.1.5 to count the number of sample medians, $\hat{y}_{med}$, that are less than 70.0. Also, count the number of sample medians greater than 110.0.

7.1.7   Calculate the sample mean, $\bar{y}$, for each of the 56 samples listed in Exercise 7.1.4. (With a hand calculator this exercise is relatively easy; without one it is somewhat tedious. Reference to Table 12.3 will help.)

7.1.8   Using the results of Exercise 7.1.7 count the number of sample means less than 75.0. Also, count the number of sample means greater than 107.0.

7.2 What Confidence Can We Have in the Sample?

The statements developed in Section 7.1 describe how far *away* the sample mean can be from the true mean; these statements can be usefully reversed. If exactly 4 of the 28 possible mean values are at least as *far away* from the true mean as 66.0 and 116.0, then the rest of the possible means, that is, 24 of 28, must be *within* the limits 66.0 and 116.0. So in advance of sampling, the probability of getting a sample with a mean value between 66.0 and 116.0 is 24/28 = 0.86. This reversed statement describes how sure we are about the closeness of the sample mean to the true mean, which is in contrast to the earlier statements about how far away it could be. This reversed statement is formally written as,

$$Prob\{66.0 < \bar{y} < 116.0\} = 24/28, \qquad (7.1)$$

which simply says that there are 24 chances out of 28 (86%) that the sample mean actually obtained will be within the limits of 66.0 and 116.0.

The statement of Equation 7.1 is still true if the interval is narrowed by one unit at each end. Specifically, if the interval {66.0,116.0} is changed to the interval {67.0,115.0}, then,

$$Prob\{67.0 < \bar{y} < 115.0\} = 24/28 . \qquad (7.2)$$

Both Equations 7.1 and 7.2 are true because the next mean larger than 66.0 is 70.0 and the next mean smaller than 116.0 is 112.0. Consequently, the count of possible sample means outside of the interval of Equation 7.2 is still 4 in 28. The reason for narrowing the interval by one unit at each end in this way is simply to avoid a problem with interval end points that would arise later when we reverse certain inequalities. There is no change in the meaning of the statement.

Equation 7.1 was written as Equation 7.2 for convenience. Now we are going to rewrite Equation 7.2 because it is informative to do so. First notice that Equation 7.2 can be rewritten in terms of differences from the population mean $\bar{Y} = 91.5$;

$$Prob\{67.0 - 91.5 < \bar{y} - \bar{Y} < 115.0 - 91.5\} = 24/28, \quad (7.3)$$

or simply,

$$Prob\{-24.5 < \bar{y} - \bar{Y} < 23.5\} = 24/28 .$$

In this form the statement says that the difference, $\bar{y} - \bar{Y}$, between the sample mean and the true mean will be greater than -24.5 and less than 23.5 with probability $24/28 = 0.86$. Mathematically trained readers will see immediately that Equation 7.3 follows from Equation 7.2, but in any case, it can easily be verified numerically by simply counting the number (24) of deviations, $\bar{y} - \bar{Y}$, which lie between -24.5 and 23.5.

Finally, Equation 7.3 can be rewritten as Equation 7.4. This has been done by subtracting $\bar{y}$ from each part of the inequality and then multiplying through by -1. The multiplication by -1 reverses the direction of inequalities. Again these steps will be readily evident to readers with some mathematical background but as we shall see Equation 7.4 can also be readily verified numerically.

$$Prob\{\bar{y} - 23.5 \leqslant \bar{Y} \leqslant \bar{y} + 24.5\} = 0.86. \quad (7.4)$$

Equation 7.4 applies *in advance of sampling;* it says that in repeated sampling intervals calculated as in Equation 7.4 will cover the true mean, $\bar{Y}$, 86% of the time. The correctness of this statement is verified by direct counting in Table 7.1. The means, $\bar{y}$, taken from Table 5.2 are listed in order of magnitude in the first column of Table 7.1. Beside each sample mean the interval associated with that mean (calculated by Equation 7.4) is listed. It clearly shows that 24 out of the 28 intervals cover the true mean 91.5, just as Equation 7.4 predicted.

Such intervals are called *confidence intervals* and the end points of the intervals are called *confidence limits*. The *size* of these confidence intervals is said to be 86%.

The interpretation of a realized confidence interval requires some care. The interval $\{68.5, 116.5\}$ appears to say that the true mean Y lies between 68.5 and 116.5 with probability 0.86. But this is clearly nonsense because in this particular case the true mean 91.5 is certainly within the interval. In this example, 0.86 is the probability that, in advance of sampling, the realized interval will cover the true mean. The confidence interval *procedure* leads to a particular *realized* confidence interval, which may or may not cover the true mean.

Confidence intervals of different size can be constructed. In the IRS example, the interval $\{70.0, 112.0\}$ contains 22 of 28 possible means and hence can be used to develop confidence intervals of size $(22/28) \times 100 = 79\%$. The size can be readily verified by counting

(Figure 5.1) that 22 of the 28 possible mean values lie within the limit: of 70.0 and 112.0.

Confidence limits need not cut off the same fraction of the sampling distribution at each end. Each of the intervals {66.0,116.0} and

TABLE 7.1

EXACT CONFIDENCE INTERVALS OF SIZE 24/28
FOR THE MEAN OF THE TAXPAYER POPULATION
(2 taxpayers from 8)

Sample Mean	Lower Limit	Upper Limit	Covers True Mean?
64.0	40.5	88.5	no
66.0	42.5	90.5	no
70.0	46.5	94.5	yes
75.0	51.5	99.5	yes
77.0	53.5	101.5	yes
79.0	55.5	103.5	yes
81.0	57.5	105.5	yes
81.0	57.5	105.5	yes
81.0	57.5	105.5	yes
81.0	57.5	105.5	yes
83.0	59.5	107.5	yes
85.0	61.5	109.5	yes
87.0	63.5	111.5	yes
88.0	64.5	112.5	yes
92.0	68.5	116.5	yes
92.0	68.5	116.5	yes
94.0	70.5	118.5	yes
95.0	71.5	119.5	yes
96.0	72.5	120.5	yes
98.0	74.5	122.5	yes
99.0	76.5	123.5	yes
101.0	77.5	125.5	yes
103.0	79.5	127.5	yes
105.0	81.5	129.5	yes
109.0	85.5	133.5	yes
110.0	86.5	134.5	yes
112.0	88.5	136.5	yes
116.0	92.5	140.5	no
123.0	99.5	145.5	no

{64.0,112.0} contains 24 of 28 possible means. The first interval has 2 of the sample means at or beyond the limits at each end, while the second interval has 1 in the lower tail and 3 in the upper tail. Both intervals can be used to develop confidence intervals of size 24/28. In the extreme, *one-sided* confidence intervals can be formed. For example,

$$Prob\{\bar{y} < 109.5\} = 24/28 = 0.86$$

so that

$$Prob\{\bar{y} - 18.0 < \bar{Y}\} = 0.86. \tag{7.5}$$

Equation 7.5 says that, in advance of sampling, the true mean $\bar{Y}$ will be greater than $\bar{y} - 18.0$ with probability 0.86. It is left to the reader to verify that this one-sided interval does indeed cover the true mean in 24 of the 28 cases.

In the two previous paragraphs different confidence intervals, all of the same size (24/28), were discussed. This leads to the question, "Among confidence intervals of the same size, which one is best to use?" A complete answer to this question requires substantial mathematical analysis and a discussion of what "best" might mean. We are not going to attempt to answer that question here. In practice, symmetric confidence intervals which cut off equal fractions of the sampling distribution at each tail are used most often. The other possibilities need not concern us now. As far as the size of the confidence interval is concerned, it is a matter of personal taste. In practice both 90 and 95% intervals are widely used.

Exercises

In the following exercises you are asked to "reverse" the statements contained Exercises 7.1.1, 7.1.2, and 7.1.3 and which are based on Table 7.1. To help you do this the "reversed" statements are given in Exercises 7.2.1 to 7.2.5.

7.2.1   How many sample means are smaller than 104.0 in Table 7.1?

7.2.2   What is the probability that in the sampling of two taxpayers from eight, the sample mean does not overestimate the true mean by more than 12.5?

7.2.3   Count the number of sample means that are larger than 78.0.

7.2.4   What is the probability that the sample mean does not under estimate the true mean by more than 13.5?

7.2.5   What is the chance that the actual sample mean will have a value between 78.0 and 104.0? An alternate wording of this question is, "What is the chance that the sample mean does not underestimate the true mean by 13.5 nor overestimate it by 12.5?"

7.2.6   Use Table 7.1 to verify Equation 7.5 in the text. To do this calculate $\bar{y} - 18.0$ for each sample by subtracting 18.0 from each of the sample means in Table 7.1. Then check that 4 of the 28 differences $\bar{y} - 18$ are greater than $\bar{Y} = 91.5$.

7.2.7   By reference to Table 7.1, numerically verify the following statements.

(a) $Prob\{\bar{y} + 13.5 \geqslant \bar{Y}\} = 23/28$

(b) $Prob\{\bar{y} - 12.5 \leqslant \bar{Y}\} = 22/28$

(c) $Prob\{-13.5 \leqslant \bar{y} - \bar{Y} \leqslant 12.5\} = 17/28.$

7.2.8   In practice, the confidence intervals used most often cut off equal amounts of the sampling distribution at each end. The confidence intervals specified in Exercise 7.2.7(c) do not do this. Numerically, verify that the confidence intervals,

$$Prob\{\bar{y} - 15.0 \leqslant \bar{Y} \leqslant \bar{y} + 13.5\}$$

cut off 5/28 of the sampling distribution at each end and hence are of size 18/28.

7.2.9   Use the counts obtained in Exercise 7.1.6 to develop a set of confidence intervals of size 44/56 for the median. (For your reference the technique required to do this exercise was used to develop the confidence intervals that appear in Equation 7.4.)

7.2.10   Use the counts obtained in Exercise 7.1.8 to obtain a set of confidence intervals for the sample mean of size 48/56.

7.2.11   Use the results of Exercises 5.2.1 and 5.2.2 to develop confidence intervals of the following types for the true mean $\bar{Y}$.

(a) Eighty percent confidence intervals which cut off equal amounts of the sampling distribution in both tails of the distribution.

(b) Eighty percent confidence intervals which cut off the entire

20% of the distribution from the upper tail of the sampling distribution.

(c) A set of 70% confidence limits.

7.3. Objectives and Obstacles

At this point, it should be clear that we want to associate confidence intervals with our sample estimates. Furthermore, the shorter these confidence intervals are the more useful they ought to be to us. A 90% confidence interval {95,105} should be better than a 90% confidence interval {90,110}. This is the reason why there was so much interest in Chapter 5 in the factors that influence the shape of the sampling distribution; these factors can be manipulated to reduce the length of confidence intervals. Most of the rest of this book is devoted to this use of these factors.

But at this point there is a major obstacle in our way. The development and examples of confidence limits in the last section used the knowledge of the *complete* sampling distribution. If this complete sampling distribution had not been available, as in Table 5.2, the confidence intervals could *not* have been constructed in this way. How then can confidence limits be constructed in practice when only a single sample is available and *not* the complete sampling distribution? It appears that confidence limits should be quite impossible to construct.

But just as mathematics showed (Section 5.5) how to estimate the variance of a sample mean from a single sample, there are even more powerful results which enable the construction of confidence limits. They are discussed in the next chapter on the truly amazing normal distribution.

IMPORTANT NEW IDEAS

upper (lower) tail of a distribution   confidence intervals

confidence limits   size of confidence intervals

CHAPTER 8

THE AMAZING NORMAL DISTRIBUTION

8.1 A Startling Theoretical Result

In the last chapter, confidence intervals for the population mean were constructed by using detailed knowledge of the sampling distribution of the sample mean. Unfortunately in practice, this distribution is *not* known in detail; in fact, only a *single* sample mean is known. So at this point, the development of confidence intervals would appear to be only a theoretical exercise. But happily, mathematical statistics has again supplied a powerful, useful result. In this case, it is the central limit theorem, one of the most amazing theorems in applied mathematics.

To observe the theorem in operation consider Figure 8.1. Figure 8.1a repeats Figure 3.2b, the distribution of forecasting errors discussed in Section 3.3. Figure 8.1b shows the histogram of 100 different random sample means, each sample of size 2, drawn from the distribution of forecasting errors. Figures 8.1c and 8.1d show similar plots of distributions of 100 sample means based on 4 and 8 observations, respectively.

Notice that the sampling distributions bunch up more in the middle as the sample size gets larger. This is easy to see because the horizontal scale in each graph is the same. The bunching up should come as no surprise, because we already know that the variance of the mean gets smaller as n gets larger (Chapter 5). This fact is being illustrated in these plots. But, as it turns out, there is another, much more fascinating feature associated with increasing sample size.

In Figure 8.1a, the population of forecasting errors has quite a skewed appearance. From Figure 8.1b, 8.1c, and 8.1d we can see that as n gets larger, the skewness almost disappears and that for $n = 8$, the histogram is reasonably well described by the smooth, symmetric, bell-shaped curve superimposed on it. This curve is the normal density which was mentioned briefly in Section 3.3.

The central limit theorem says (loosely) that as the sample size gets larger, the sampling distribution of the mean is better and better approximated by the normal distribution. The central limit property is surprisingly general. It requires mainly that as n gets larger so does N and $N - n$. The theorem does *not* require that the sampled population have any particular shape; the result usually holds. The population of forecast errors (Figure 8.1a) is longer on the right than on the left, and is not particularly bell-shaped. Nevertheless, the sampling distribution of means from this non-normal population *does* begin to look normal as n increases. There are a few restrictions on the generality of the theorem which are referred to later in this chapter, but their precise mathematical nature need not concern us in this book.

If the sampled population is itself a normal population, the central limit theorem is not approximate; it is exact. That is, the sampling distribution of the mean is exactly a normal distribution, for every value of n including small values. On the other hand, if the shape of the population is very non-normal, say with one tail much longer than the other, a larger sample size will be needed for the symmetric normal distribution to appear. For the most common populations, samples of size 25 to 30 do surprisingly well at inducing good central limit approximations. In most real studies, the samples are much larger than 30. As a result, the theorem is used a great deal in practice.

In summary, we know from Chapter 5 that the sampling distribution of the mean $\bar{y}$ has a mean value equal to $\bar{Y}$, the true population mean. We also know from Chapter 5 that the sampling variance of $\bar{y}$ for large populations is σ^2/n. If these two facts are combined with the central limit theorem, we know that the sampling distribution of $\bar{y}$, for reasonably large values of n, is normal with mean $\bar{Y}$ and variance σ^2/n. This is an extremely handy approximation.

Exercises

8.1.1   Use the intervals [60,70), [70,80), [80,90), [90,100), [100,110), [110,120), [120,130) to plot the actual incomes of the 8 taxpayers as a histogram (see Section 3.3). In the notation above the curved and square brackets imply that the interval [70,80) includes the point 70 and all points between 70 and 80, but *not* including the specific point 80. Similar interpretations apply to the other intervals.

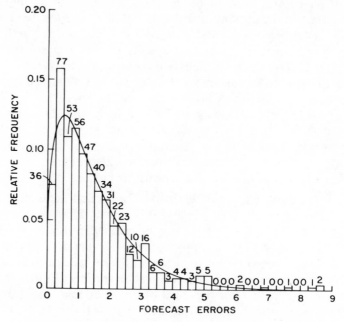

Figure 8.1a. Distribution of corporate forecast errors.

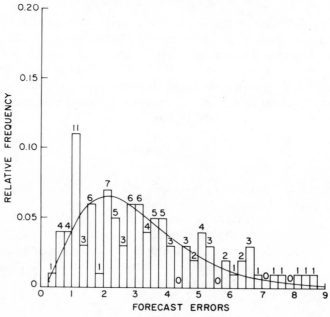

Figure 8.1b. Distribution of 100 sample means, $n = 2$.

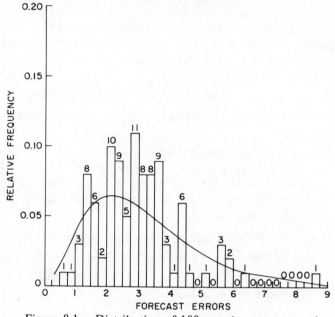

Figure 8.1c. Distribution of 100 sample means, $n = 4$.

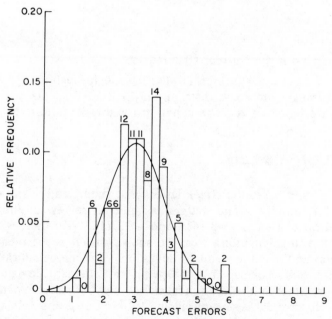

Figure 8.1d. Distribution of 100 sample means, $n = 8$.

8.1.2   Use the intervals given in Exercise 8.1.1 to plot the sampling distribution of $\bar{y}$ (Table 5.2) as a histogram rather than the line graph given in Figure 5.1. Does the comparison of the histograms obtained in Exercises 8.1.1 and 8.1.2 show any of the characteristics associated with the central limit theorem? If so, describe them.

8.1.3   The number of possible samples of size 4 from a population of size 8 is $\binom{8}{4} = 70$. So even with most hand calculators, the calculation of all possible sample means is tedious. To save some effort, pair the 28 sample means in Table 5.2 to make 14 samples each based on 4 observations. Plot these 14 means using the intervals given in Exercise 8.1.1. Next, are the central limit characteristics any more evident than they were in Exercise 8.1.2? Describe how.

8.1.4   Does the method of pairing samples in Exercise 8.1.3 yield a subset of the $\binom{8}{4} = 70$ different samples that are possible by selecting 4 from 8 with equal probability without replacement? Give reasons for your answer.

8.2 Properties of the Normal Distribution

We encountered the normal distribution in Sections 3.3 and 8.1. It is the most important distribution in statistics; many statistical methods are based on it. The equation for this symmetric bell-shaped curve is,

$$f(X) = \frac{1}{\sigma \sqrt{2\pi}} \, e^{-\frac{1}{2}\left(\frac{X-\mu}{\sigma}\right)^2}, \quad -\infty < X < \infty \qquad (8.1)$$

where μ, σ, and π are the Greek letters mu, sigma, and pi and ∞ is the symbol for infinity. The letters π and e represent the known mathematical constants $\pi = 3.1416$ and $e = 2.7183$. Further, with some mathematical manipulation, it can be shown that μ is the mean of the distribution and σ is the standard deviation. The mathematically sensitive reader will recognize that the mean and standard deviations of a continuous, smooth distribution such as the normal, have not been defined. However, it is not misleading to think of μ and σ in the same way that the mean and standard deviation were discussed for finite populations.

The normal distribution has the property that changing from X to $Z = (X - \mu)/\sigma$, always results in Z having a normal distribution with mean 0 and variance 1, that is,

$$f(Z) = \frac{1}{\sqrt{2\pi}} e^{-\frac{Z^2}{2}}, \quad -\infty < Z < \infty . \qquad (8.2)$$

This is a very helpful form of the normal distribution because the formula does not contain any unknown parameters, such as μ and σ. As a result, it is easily tabulated, and we shall make good use of this property.

The total area under the normal curve and above the x axis is scaled to be equal to one. In fact, this is the sole purpose of the divisor $\sigma \sqrt{2\pi}$ in Equation 8.1. This scaling is done for the same reasons of convenient interpretation that were discussed in Section 3.3. Consequently, the area under the curve to the left (right) of a specific point $Z = a$ represents the proportion of cases in the sampling distribution that are less (greater) than a. Such points are called *quantiles* of the distribution. The area *between* the two quantiles, $Z = b$ and $Z = a$, where b is greater than a, represents the chance of getting a sample value less than b but greater than a. Since this is exactly the type of information that was used in Chapter 7 for the construction of confidence intervals, it is of great potential use to be able to determine the area under sections of the normal distribution. That is what we are going to discuss in this section.

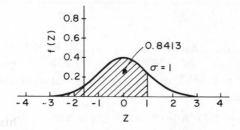

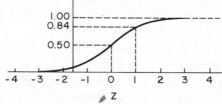

Figure 8.2. The cumulative normal distribution.

The *cumulative* normal distribution gives the relative frequency with which observations fall *below* particular points. For example, in Figure 8.2 the top graph is the normal distribution with mean 0 and variance 1. It has been calculated elsewhere that the shaded area to the left of $Z = 1$ represents 0.8413 of the total area under the curve. In the cumulative normal distribution the *height* of ordinate at the point $Z = 1$ is equal to 0.8413; the area under the normal curve up to the point $Z = 1$. The cumulative normal distribution appears as the bottom plot in Figure 8.2.

As remarked earlier, the normal distribution is easily tabulated in the standardized form with mean 0 and variance 1. As we shall see, the tabulation of the cumulative normal distribution is the handiest to use. Some values of Z and the associated cumulative areas are given in Table 8.1. More detailed tables are readily available in the statistics literature.

Since the cumulative normal distribution gives the proportion of area *below* specified points, the proportion *between* two points is found by subtraction. To illustrate, in a normal distribution with $\mu = 40$ and $\sigma = 5$, what fraction of the area falls between 30 and 45? To answer this, first calculate the specific Z values as follows,

$$Z_1 = \frac{30 - 40}{5} = -2.0$$

and

$$Z_2 = \frac{45 - 40}{5} = +1.0.$$

Next Table 8.1 shows that the area to the left of $+1.0$ is 0.8413 and the area to the left of -2.0 is 0.0228. This means that $0.8413 - 0.0228 = 0.8185$ of the area lies between the points 30 and 45. See Figure 8.3.

The normal distribution is of great use in the construction of confidence intervals. This is done in Section 8.3. In that section it will be helpful if we know how to answer questions of the type, "In the normal distribution with $\mu = 40$ and $\sigma = 5$, what two values contain the *central* 95% of all area under the curve?"

Since the central area is wanted, 0.025 or 2-1/2% of the area must be cut off from each tail. From Table 8.1, we see by interpolating between the points -1.90 and -2.00 and the points 1.90 and 2.00 that 0.025 of the area is cut off at the point -1.96 in the lower tail and $+1.96$ in the upper tail. Consequently, solving

TABLE 8.1

THE CUMULATIVE NORMAL DISTRIBUTION

Z	Area	Z	Area
−3.20	0.0007	0.00	0.5000
−3.10	0.0010	0.10	0.5398
−3.00	0.0013	0.20	0.5793
−2.90	0.0019	0.30	0.6179
−2.80	0.0026	0.40	0.6554
−2.70	0.0035	0.50	0.6915
−2.60	0.0047	0.60	0.7257
−2.50	0.0062	0.70	0.7580
−2.40	0.0082	0.80	0.7881
−2.30	0.0107	0.90	0.8159
−2.20	0.0139	1.00	0.8413
−2.10	0.0179	1.10	0.8643
−2.00	0.0228	1.20	0.8849
−1.90	0.0287	1.30	0.9032
−1.80	0.0359	1.40	0.9192
−1.70	0.0446	1.50	0.9332
−1.60	0.0548	1.60	0.9452
−1.50	0.0668	1.70	0.9554
−1.40	0.0808	1.80	0.9641
−1.30	0.0968	1.90	0.9713
−1.20	0.1151	2.00	0.9772
−1.10	0.1357	2.10	0.9821
−1.00	0.1587	2.20	0.9861
−0.90	0.1841	2.30	0.9893
−0.80	0.2119	2.40	0.9918
−0.70	0.2420	2.50	0.9938
−0.60	0.2743	2.60	0.9953
−0.50	0.3085	2.70	0.9965
−0.40	0.3446	2.80	0.9974
−0.30	0.3821	2.90	0.9981
−0.20	0.4207	3.00	0.9987
−0.10	0.4602	3.10	0.9990
0.00	0.5000	3.20	0.9993

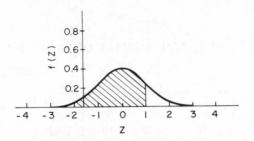

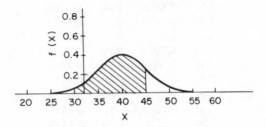

Figure 8.3. Areas of the normal distribution.

$$Z_1 = -1.96 = \frac{X_1 - 40}{5}$$

and

$$Z_2 = +1.96 = \frac{X_2 - 40}{5}$$

gives $X_1 = 30.2$ and $X_2 = 49.8$ as the two points which contain the central 95% of the area.

Exercises

8.2.1   Find the area in the normal curve to the left of the following
        Z values. $Z:\{-3.10, \ -2.30, \ -1.50, \ +0.50, \ +1.50, \ +1.80,$
        $+2.70\}$.

8.2.2   Find the area in the normal curve to the right of the Z values given in Exercise 8.2.1.

8.2.3   What fraction of the area in the normal curve lies between the following pairs of Z values?

$$
\begin{array}{rcr}
-3.10 & \text{and} & -1.50 \\
-2.30 & \text{and} & +2.30 \\
-1.50 & \text{and} & +0.50 \\
-1.50 & \text{and} & +1.80 \\
+0.50 & \text{and} & +2.70 \\
\end{array}
$$

8.2.4   Find the Z values that cut off the following percentages in the left tail of the normal distribution, $P:\{2.28, 9.67, 42.07, 50.00, 90.32, 97.72\}$.

8.2.5   Find the Z values that cut off the percentages of area given in Exercise 8.2.4 in the *right* tail of the normal distribution.

8.2.6   Find the Z values that cut off 5, 10, and 25 percent of the area in the left tail of the normal distribution.

The answers to this exercise are not directly given in Table 8.1. Interpolation is required. For example, the point $Z = -2.20$ cuts off 13.9% of the area and the point $Z = -2.10$ cuts of 17.9% of the area. So the point that cuts of 15% of the area is 1.1/4.0, or just a little more than 1/4 of the way from -2.20 to -2.10. Therefore, the point -2.17 cuts off about 15% of the area in the left tail of the normal distribution.

8.2.7   Find the Z values that cut off the 5, 10, and 25% of the area in the right tail of the normal distribution.

8.2.8   Find the two Z values that contain the *central* 90% of the area of the normal distribution. Similarly, find the Z values that contain the central 20% and the central 50%.

8.3 Normal Confidence Limits

In Chapter 7, the possibility of actually obtaining confidence intervals was blocked by the fact that the sampling distribution of the mean was unknown. But now, the way has been cleared by the central limit theorem which states that, in a very wide variety of situations, the sampling distribution of the mean is approximately normal. Specifically, the sampling distribution of $\bar{y}$ is approximately normal with mean $\bar{Y}$ and variance σ^2/n. Also, from the last section, we know that, the ratio,

$$\frac{\bar{y} - \bar{Y}}{\sigma/\sqrt{n}} \tag{8.3}$$

is approximately normal with mean 0 and variance 1. Consequently, if we knew σ, confidence intervals could be constructed for $\bar{Y}$ in exactly the way described in Chapter 7. But one difficulty remains, in most real problems, σ is not known. To construct confidence limits in practice we must deal with this one last obstacle.

Fortunately, mathematical statistics has again come to the rescue by showing that if the sample value, s, is substituted for σ, the normal approximation is still useful. That is, the sampling distribution of,

$$\frac{\bar{y} - \bar{Y}}{s/\sqrt{n}} \tag{8.4}$$

is approximately normal with mean 0 and variance 1. A slightly better approximation is given by a distribution called the "t" distribution. But for our purposes we can ignore this because the normal distribution and the t distribution are almost the same for samples of size 30 and over. For smaller values of n a slightly improved normal approximation is obtained by replacing the divisor of $n-1$ in s by the divisor $n+2$. But for our present purposes even this can be ignored. The substitution of the sample estimate s for the unknown σ removes the last block to the construction of approximate confidence intervals.

So, to obtain 95% normal confidence intervals proceed as follows. From the central limit theorem and the various approximations discussed, we know that $\sqrt{n}\,(\bar{y}-\bar{Y})/s$ can reasonably be assumed to have a normal distribution with mean 0 and variance 1. Consequently, from the tabulation of the normal distribution, Table 8.1,

$$Prob\left\{-1.96 \leqslant \frac{\bar{y} - \bar{Y}}{s/\sqrt{n}} \leqslant 1.96\right\} = 0.95.$$

These inequalities can be manipulated just as in Chapter 7 to give

$$\left\{\bar{y} - 1.96\,\frac{s}{\sqrt{n}},\ \bar{y} + 1.96\,\frac{s}{\sqrt{n}}\right\} \tag{8.5}$$

as 95% confidence intervals for the true mean $\bar{Y}$.

There is no reason why the size of the confidence intervals must be 95%. To illustrate, from Table 8.1 it can be seen that 0.05 of the area lies to the left of the point -1.65 (about 1/2 way between -1.60 and -1.70). Similarly, 0.95 of the area lies to the left of the point $+1.65$. So by subtraction 0.05 of the area lies to the right of the point $Z = 1.65$. Consequently,

$$Prob\{-1.65 \leqslant \frac{\bar{y} - \bar{Y}}{s/\sqrt{n}} \leqslant 1.65\} = 0.90,$$

and so

$$\{\bar{y} - 1.65 \frac{s}{\sqrt{n}}, \ \bar{y} + 1.65 \frac{s}{\sqrt{n}}\}$$

are 90% confidence intervals for $\bar{Y}$.

In general notation, the size of the confidence interval is usually denoted by $1 - \alpha$ where α is the Greek letter alpha. Also, $Z_{\alpha/2}$ and $Z_{1-\alpha/2}$ are the quantiles which cut off $\frac{\alpha}{2}$ of the area in each tail of the distribution. Then, by exactly the same manipulation as previously,

$$\{\bar{y} + Z_{\alpha/2} \frac{s}{\sqrt{n}}, \ \bar{y} + Z_{1-\alpha/2} \frac{s}{\sqrt{n}}\} \qquad (8.6)$$

are $(1 - \alpha)\%$ confidence intervals for $\bar{Y}$.

While the central limit theorem has quite broad applicability, it should not really be used for the construction of confidence limits in cases like the IRS sampling study. The sampled population should be smooth, and not small and discrete like the population of eight taxpayers. In addition, the sample $(n=2)$ is much too small to expect the central limit approximations to be useful. But let's apply the method anyway.

If sample 16 is selected, then $\bar{y} = 85,00$, $s^2 = 578$, and the approximate 95% confidence limits are (using Expression 8.5),

$$\{85.00 - 1.96 \frac{\sqrt{578}}{\sqrt{2}}, \ 85.00 + 1.96 \frac{\sqrt{578}}{\sqrt{2}}\}$$

or upon arithmetic simplification,

$$\{51.68, \ 118.32\}.$$

This calculation has been made for *all* the possible samples in the IRS study. The approximate normal confidence limits for the 28 samples are presented in Table 8.2. Five of the 28 intervals do *not* cover the true mean, so that the actual size of the confidence interval is $23/28 = 0.82$ rather than 0.95, as indicated by the central limit formulation. In view of the small size of the sample, and the small size of the population, it is remarkable that the results are as close as they are.

Finally, we point out in advance that this same normal distribution methodology is useful for the other estimators which are discussed later in the book. In fact, if θ is an unknown parameter, $\hat{\theta}$ is an estimate of θ, and $var(\hat{\theta})$ is an estimate of $Var(\hat{\theta})$, then

$$\{\hat{\theta} - Z_{\alpha/2}\sqrt{var(\hat{\theta})}, \ \hat{\theta} + Z_{1-\alpha/2}\sqrt{var(\hat{\theta})}\}. \tag{8.7}$$

is an approximate $(1-\alpha)\%$ confidence interval for θ. Here θ is the Greek letter theta, which is commonly used in statistical literature to represent unknown population parameters. It is also customary to use "theta hat," $\hat{\theta}$, to represent an estimate of θ.

Exercises

8.3.1   If a sample of size 20 has a mean $\bar{y} = 40$ and standard deviation $s = 5$, calculate an approximate 95% confidence interval for the true mean. For the same mean and standard deviation, calculate an approximate 80% confidence interval.

8.3.2   If a sample of size 36 has a mean of $\bar{y} = 50$ and a standard deviation of $s = 10$, calculate an approximate 90% confidence interval for the true mean. For the same mean and standard deviation, calculate an approximate 50% confidence interval.

8.3.3   A sample of size 10 was drawn from a large population and the following observations made:

$$5, 10, 6, 8, 4, 8, 6, 6, 5, 8.$$

(a) What statements can be made about the population mean?
(b) What assumptions are involved in making these statements?

8.3.4   In a school system the average IQ of a sample of 50 students was found to be $\bar{y} = 108$ with $s = 11.3$. Find a 95% confidence interval for the true mean.

8.3.5   In a second school system the average IQ of a sample of 25 students was found to be $\bar{y} = 102$ with $s = 15$. Find a 95% confidence interval for the true mean.

8.3.6   On the basis of Exercises 8.3.5 and 8.3.6 would you think that on the average the students in the second school system clearly have lower IQ's?

TABLE 8.2

APPROXIMATE NORMAL CONFIDENCE LIMITS
FOR THE TRUE MEAN OF THE TAXPAYER POPULATION
(2 taxpayers from 8)

Sample Number	Sample Mean	Lower Limit	Upper Limit	Contains True Mean?
1	66.0	54.2	77.8	no
2	64.0	56.2	71.4	no
3	77.0	43.7	110.3	yes
4	75.0	45.6	104.4	yes
5	81.0	39.8	122.2	yes
6	88.0	33.1	142.9	yes
7	95.0	26.4	163.6	yes
8	70.0	66.1	73.9	no
9	83.0	61.4	104.6	yes
10	81.0	63.4	98.6	yes
11	87.0	57.6	116.4	yes
12	94.0	50.9	137.1	yes
13	101.0	44.2	157.8	yes
14	81.0	55.5	106.5	yes
15	79.0	57.4	100.6	yes
16	85.0	51.7	118.3	yes
17	92.0	45.0	139.0	yes
18	99.0	38.2	159.8	yes
19	92.0	88.1	95.9	yes
20	98.0	90.2	105.8	yes
21	105.0	83.4	126.6	yes
22	112.0	76.2	147.3	yes
23	96.0	84.2	107.8	yes
24	103.0	77.2	128.5	yes
25	110.0	70.8	149.2	yes
26	109.0	95.1	122.7	no
27	116.0	88.6	143.4	yes
28	123.0	109.3	136.7	no

8.3.7 The percentage rate for various candidates in the 1948 Presidential election is given below. The Gallup poll predictions are also given.

	Truman	Dewey	Wallace	Thurmund	Other
Election	49.5	45.1	2.4	2.4	0.6
Gallup	44.5	49.5	4.0	2.0	0.0

Calculate an approximate 95% confidence interval for Truman's percentage vote. Assume that the sample size was 2000.

IMPORTANT NEW IDEAS

normal distribution

quantiles

central limit theorem

CHAPTER 9

THE ULTIMATE OBJECTIVE

9.1 Estimates That Are Close to the Target

Our ultimate objective is the development of estimates that are as close as possible to the true mean of the target population. We know now that both bias and variance affect this closeness, but so far in this book, these two factors have been discussed separately. The most common way to combine variance and bias is in the mean square error.

The mean square error has a simple logic. To discuss it, suppose that in general we have an estimator $\hat{\theta}$ of a true population value[1] θ. For example in a specific case which is familiar to us, the sample mean $\bar{y}$ is an estimator of $\bar{Y}$. However in general, suppose that $\hat{\theta}$ is biased, $E(\hat{\theta}) \neq \theta$, that is, the average value of $\hat{\theta}$ over all the possible values is not equal to θ. Then the error made by $\hat{\theta}$ as an estimator of θ is $\hat{\theta} - \theta$, and further, $E[\hat{\theta} - \theta]^2$ is the average of all the possible *squared* errors. The squared error is used for the same reasons of convenience that squared deviations are used in the variance, (see Chapter 3). This average squared error is called the mean square error of $\hat{\theta}$, $MSE(\hat{\theta}) = E[\hat{\theta} - \theta]^2$.

The mean square error can be conveniently and informatively rewritten by some very simple mathematical steps, which the reader may wish to try for himself,

$$E[\hat{\theta} - \theta]^2 = E[\hat{\theta} - E(\hat{\theta}) + E(\hat{\theta}) - \theta]^2$$
$$= E[\hat{\theta} - E(\hat{\theta})]^2 + [E(\hat{\theta}) - \theta]^2 . \qquad (9.1)$$

The first step in Expression 9.1 is an easy one. In this step, the deviation of the estimate from the true value, $\hat{\theta} - \theta$, is separated into two components, $\hat{\theta} - E(\hat{\theta})$ and $E(\hat{\theta}) - \theta$ by the simple subtraction and addition of $E(\hat{\theta})$. The first component is the deviation of the

---

1. As in Chapter 8, θ is the Greek letter "theta," which is commonly used in statistics to represent population parameters. It is also common to use "theta hat," $\hat{\theta}$, to represent an estimator of θ.

estimate from its expected value. The expectation of the squares of these deviations is simply the variance of $\hat{\theta}$,

$$E[\hat{\theta} - E(\hat{\theta})]^2 = Var(\hat{\theta}).$$

Furthermore, the second component $E(\hat{\theta}) - \theta$, is the bias in $\hat{\theta}$. Consequently, we can see that Equation 9.1 says that

$$MSE(\hat{\theta}) = Var(\hat{\theta}) + [Bias\ in\ \hat{\theta}]^2. \tag{9.2}$$

Clearly, if an estimator is unbiased, its mean square error and variance are identical.

It needs to be pointed out that while the first step in Expression 9.1 is an easy one for almost all readers, the second step may not be. It requires the observation that the cross-product term, $E(\hat{\theta} - E(\hat{\theta}))(E(\hat{\theta} - \theta)) = 0$. However, this difficulty is not a hindrance to the reader because all that he really needs to understand is that the mean square error can be written in terms of the two components bias and variance, as indicated in Equation 9.2.

The fact that the mean square error can be written in terms of bias and variance is a major reason for its appeal. Bias and variance are quantities with which we are already familiar and which are important for their own unique characteristics. A second reason is that sometimes it is handy to calculate the mean square error of an estimator by calculating the bias and variance separately. This permits two estimators to be compared directly on the basis of mean square error. Such comparisons are made in Chapter 12. As we shall see, by permitting a small amount of technical bias, variance can sometimes be reduced substantially; this implies a much smaller mean square error.

To illustrate the mean square error in a different way, think of the sampling scheme as a rifle shot repeatedly at the center of a target. If the rifle is aimed directly at the center, C, the shots will be spread randomly around C. That's variance. On the other hand, if the rifle is *not* aimed at the center of the target (or if there is a systematic effect such as wind or a defect in the rifle) but rather at some other point, A, then the shots will be spread around in the vicinity of the point A. This spread around the point A corresponds to variance and the distance $A - C$ to bias.

Clearly, to get as close to the true parameter as possible both bias and variance must be minimized. The discussion of variance reduction begins in the next section. Measurement biases were discussed in Section 6.2, selection biases are discussed in Chapter 10 and technical biases in Chapter 12.

Exercises

9.1.1   What is the difference between the mean square error and variance if the estimator in question is unbiased?

9.1.2   It is sometimes argued that variance is more important than mean square error because analysts should always use estimators which are unbiased. Discuss this statement.

9.1.3   Verify the algebra in Equation 9.1. That is, show that

$$MSE(\hat{\theta}) = Var(\hat{\theta}) + (Bias(\hat{\theta}))^2.$$

9.2 Reducing Variance

The goal of estimates that are as close to the center of the target as possible implies that the sampling variance be as small as possible. Furthermore, since the sampling distribution of the mean depends upon (1) the sample size, (2) the estimator, (3) the method of sample selection, and (4) the parent population, it is these factors that can be manipulated to reduce variance. How to do this is described in the next four chapters, which is in fact most of the rest of the book. But while these factors are discussed in separate chapters, it is important to remember that they are interrelated. As we see in Chapter 11, poor coordination of the selection procedures and the natural grouping of the target population can have disastrous results.

Finally it is important to remark that the ultimate judgement of "closeness" must be made in the light of the particular use for the estimates. There is no practical interest in a 90% confidence interval which states that the average number of children per family is between 0 and 8. Even a poor guess would be within that range, and such a wide confidence interval has no practical value.

9.3 Confidence Intervals Do Not Account for Bias!

Confidence intervals measure sampling variability, *not* bias. It is possible for 95% confidence limits to hardly ever cover the true mean, never mind the supposed 95% of the time! The cause of the problem is *bias,* and it is very important to keep this fact in mind. As always the point seems best made by an example.

Suppose in the IRS example, that a computer mistake subtracts 20,000 dollars from each of the incomes. This is measurement error of the type discussed earlier (6.2); however, biases that are discussed later in the book have the same misleading potential. What effect would such an error have on the 95% normal confidence intervals developed in Chapter 8? Recall that the intervals are of the form,

$$\{\bar{y} - 1.96 \frac{s}{\sqrt{2}} , \ \bar{y} + 1.96 \frac{s}{\sqrt{2}}\},$$

where

$\bar{y}$ is the sample mean,

s^2 is the sample variance,

$n = 2$ is the sample size, and

$-1.96, 1.96$ are quantiles of the normal distribution.

To determine the effect of the numerical error, we ask which of the factors involved in the calculation of the confidence limits are changed by the numerical error and which are not?

Clearly, the numerical error reduces the sample mean by 20. But just as clearly, $n = 2$ and the normal points -1.96 and 1.96 are unchanged. Furthermore, s^2 is also unchanged. This can be seen by considering the sum of squares of deviations in the numerator of s^2. Each new deviation, based on the faulty measurement, is equal to $(y_i -20) - (\bar{y} -20) = y_i - \bar{y}$, that is, the deviations are unchanged. Of course, the sum of the squares of the deviations is also unchanged. Finally, since the denominator, $n-1$, is unchanged, s^2 is completely unchanged by the error. So in total, all the numerical factors that go into the confidence interval are unaffected by the error except $\bar{y}$, which is smaller by 20. Unfortunately this has the direct affect of shifting the confidence interval 20 units to the left on the income scale. In terms of the correct mean, $\bar{y}$, the biased 95% confidence intervals are now

$$\{\bar{y} - 20.0 - 1.96 \frac{s}{\sqrt{2}}, \ \bar{y} - 20.0 + 1.96 \frac{s}{\sqrt{2}}\},$$

that is, the upper and lower confidence limits of Table 8.2 are both reduced by 20. The reader can easily check in Table 8.2 that these intervals now cover the true mean in only $16/28 = 0.57$ of the cases. Thus the "95% confidence intervals" are actually 57% confidence intervals! Furthermore, if the numerical error was bigger than $163.6 - 91.5 = 72.1$, *none* of the "95%" confidence intervals would cover the true mean of 91.5. This can be verified by observing that for all the possible samples the largest of the upper limits is 163.5. If the error were large enough to shift this upper limit below the value of 91.5, then *none* of the intervals would cover the true mean. A bad situation indeed.

It is important to recall that the preceding "95%" confidence intervals are based on the central limit normal approximation to the sampling distribution. As we saw in Section 8.3, this approximation results in the "95%" confidence intervals actually being of size $(23/28)100 = 82\%$ rather than the stated 95%. Consequently, it would be more accurate in the preceding paragraph to say that the confidence intervals have been reduced in size from 82% to 57%. The point of example is unchanged, however; confidence limits do not account for bias.

The conclusion of this section is *very* important. Make a substantial effort to reduce *both* bias and variance separately, and don't be lulled into a false sense of security by narrow confidence limits when there has been no accounting for bias.

For example, in the Monday July 21, 1975, New York Times, the results of a survey on peoples' attitudes towards the fiscal plight of New York City were presented. In the statistical description of the survey, the following statement was made. "A total of 420 persons were interviewed, a random sample that statistical experts say yields 95 percent confidence that the results are within 5 percentage points of the attitudes of the population as a whole." But no statement about possible bias was made. As a result, we can only guess about the actual correctness of the confidence interval statement. In subsequent reports on other surveys, the New York Times did in fact remark on this ever-present possibility of bias and its effect on confidence limits, although they gave little reason for the reader to conclude that the bias was either large or small.

Exercises

9.3.1   Review your answer to Exercise 9.1.2 in the light of the discussion of Section 9.3. Does your revised answer differ in any significant way from your answer to Exercise 9.1.2?

9.3.2   Educated "guesses" are sometimes useful in statistics. Suppose we have information that in the reduced IRS taxpayer population, Table 5.1, that the true mean of the actual incomes is "about" 95. To take advantage of this information, it is decided that 10 should be added to every observation below 95 and 10 subtracted from every observation above it. Then the sample mean is calculated on these new numbers. The hope is that the sampling distribution of the sample mean calculated in this way will have a variance which is less than the sample mean based on the original numbers. For this "adjusted" estimator calculate, its (a) expectation, (b) bias, (c), variance, (d) mean square error.

9.3.3   Compare the mean square error of the estimator developed in Exercise 9.3.2 with the mean square error of the sample mean discussed in Chapter 5. What are the implications of your finding?

9.3.4   Find both exact and approximate confidence intervals for the estimator in Exercise 9.3.2. Select an appropriate size for your confidence limits.

9.3.5   If in Exercise 9.3.2, a numerical error is made, and 10 is mistakenly subtracted from *every* observation, what is the effect on the confidence limits calculated in Exercise 9.3.4?

IMPORTANT NEW IDEAS

mean square error

CHAPTER 10

GATHERING UP THE SAMPLE

Earlier in the book (9.2), it was stated that the method of sample selection is one of the factors that influences the sampling distribution. The implication is that sampling procedures exist which reduce the variance of the sampling distribution. But so far no such procedures have been discussed. This discussion begins in this chapter and is continued in the next.

Sampling *without* replacement with equal probability of selection is the only method of sample selection that we have discussed so far (Chapter 5). In Section 10.1, sampling with equal probability *with* replacement is discussed. It turns out that sampling with replacement and sampling without replacement are very much alike when the sampling fraction, n/N, is small. In practice, most sampling studies have a small sampling fraction. However, sampling with replacement with equal probability leads smoothly into a discussion (10.2) of sampling with replacement with *unequal* probability which, as we shall see, can have a major effect on the characteristics of the sampling scheme. Sampling with unequal probability can be much better than sampling with equal probability. And it can be worse too. Some of the difficulties are discussed in Section 10.3.

Finally, systematic sampling is discussed in Section 10.4. This procedure is not pursued extensively because systematic sampling can lead to very complicated analysis. However, there are some easy results which can be discussed and which are sufficient to make systematic sampling extremely useful in some applied situations.

10.1 Sampling with Replacement

10.1.1 What Is It?

Recall from Chapter 5, that if eight numbered discs, one for each of the eight taxpayers, are placed face down on a table and thoroughly mixed, then a sample of size two can be selected by first randomly picking one disc and then, *without* replacing the selected disc, randomly selecting a second one. This is sampling *without* replacement. On the other hand, if the disc selected first is *replaced* on the table *before* the random selection of a second disc, then the sampling is said to be *with* replacement.

TABLE 10.1

WITH REPLACEMENT SAMPLING OF TAXPAYERS

[2 taxpayers from 8]
[64 equally likely ordered pairs]
[Unit listed first is selected first]

Taxpayers Selected	Sample Mean	Taxpayers Selected	Sample Mean	Taxpayers Selected	Sample Mean	Taxpayers Selected	Sample Mean
1,1	60.0	2,1	66.0	3,1	64.0	4,1	77.0
1,2	66.0	2,2	72.0	3,2	70.0	4,2	83.0
1,3	64.0	2,3	70.0	3,3	68.0	4,3	81.0
1,4	77.0	2,4	83.0	3,4	81.0	4,4	94.0
1,5	75.0	2,5	81.0	3,5	79.0	4,5	92.0
1,6	81.0	2,6	87.0	3,6	85.0	4,6	98.0
1,7	88.0	2,7	94.0	3,7	92.0	4,7	105.0
1,8	95.0	2,8	101.0	3,8	99.0	4,8	112.0

Taxpayers Selected	Sample Mean	Taxpayers Selected	Sample Mean	Taxpayers Selected	Sample Mean	Taxpayers Selected	Sample Mean
5,1	75.0	6,1	81.0	7,1	88.0	8,1	95.0
5,2	81.0	6,2	87.0	7,2	94.0	8,2	101.0
5,3	79.0	6,3	85.0	7,3	92.0	8,3	99.0
5,4	92.0	6,4	98.0	7,4	105.0	8,4	112.0
5,5	90.0	6,5	96.0	7,5	103.0	8,5	110.0
5,6	96.0	6,6	102.0	7,6	109.0	8,6	116.0
5,7	103.0	6,7	109.0	7,7	116.0	8,7	123.0
5,8	110.0	6,8	116.0	7,8	123.0	8,8	130.0

$$E(\bar{y}) = 91.5 = \bar{Y}$$

$$E(\bar{y}-\bar{Y})^2 = \frac{1}{64}[(60-91.5)^2 + (66-91.5)^2 + \cdots + (130-91.5)^2]$$

$$= 257.88 = \frac{515.75}{2} = \frac{\sigma^2}{n}$$

Clearly, sampling with and without replacement are different. For one thing, sampling with replacement permits duplicates to appear in the same sample. To illustrate, it is possible to select disc number 3 on *both* the first and second draw, and thereby get a sample of $\{y_3 = 68.0, \ y_3 = 68.0\}$. Sampling without replacement does not permit such samples. Since we have placed great importance on the sampling distribution, an immediate question is whether and how this possibility of duplicates affects the sampling distribution of the sample mean, and in particular its mean and variance.

10.1.2 The Sampling Distribution

The new sampling distribution comprises all the possible values of the sample mean that can arise by sampling with replacement. How many such samples are there? In the taxpayer example, there are eight discs that can be selected on the first draw, and any of the same eight can be selected again on the second draw. This means that there are 8×8 equally likely possibilities; all 64 are listed in Table 10.1.

Each of the 64 possible outcomes listed in Table 10.1 is equally likely to arise, that is, be the one that is actually selected. Furthermore, one of these samples *must* occur, no others are possible. This set of 64 sample means is the with replacement sampling distribution of $\bar{y}$, it is plotted in Figure 10.1.

As with any distribution, this sampling distribution has a mean and variance. The mean of the 64 possible sample means, which we have also called the expectation of $\bar{y}$, $E(\bar{y})$, has been calculated (Table 10.1) to be equal to 91.5. This is exactly equal to the mean of the target population. So, just as we saw earlier when sampling without replacement, $\bar{y}$ is also an unbiased estimator of $\bar{Y}$ when sampling with replacement. This is again a general result which can be proven mathematically. So, the sample mean is unbiased when sampling is either with or without replacement.

But the variance of this new sampling distribution is *not* the same as when sampling is without replacement. The with replacement variance is larger, 257.88 (Table 10.1) compared to 221.04 (Table 5.2), and it is not hard to see why. Sampling with replacement permits the possible selection of the sample $\{y_1 = 60.0, y_1 = 60.0\}$ with mean $\bar{y} = 60.00$; a sample mean this small cannot occur when sampling without replacement. Similarly, the sample $\{y_8 = 130.0, y_8 = 130.0\}$ has a mean $\bar{y} = 130.00$ which is larger than any sample mean possible when sampling without replacement. Thus the range of possible sample means has increased from $123.0 - 64.0 = 59.0$ for sampling without

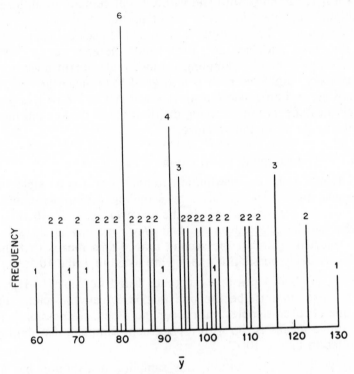

Figure 10.1. With replacement sampling distribution of $\bar{y}$
(two taxpayers from eight).

replacement to $130.0 - 60 = 70.0$ for sampling with replacement. In total, the occurrence of duplicates tends to spread the new sampling distribution out. This characteristic is reflected in the increased variance of the sampling distribution.

At the bottom of Table 10.1 it is shown numerically that the variance of the sampling distribution is equal to σ^2/n. This is another general result; it can be proven mathematically that when sampling is with replacement the variance of sample mean is indeed given by $Var(\bar{y}) = \sigma^2/n$.

The variance of the mean when sampling with replacement can be compared by formula with the variance of the mean when sampling without replacement, Specifically, the formula,

$$Var(\bar{y}) = \frac{\sigma^2}{n}, \qquad (10.1)$$

for sampling with replacement, can be compared to the formula,

$$Var(\bar{y}) = (1 - \frac{n}{N}) \frac{S^2}{n} \tag{10.2}$$

for sampling without replacement. First notice that the finite population correction does not appear in the with replacement formula. This is logical because sampling with replacement can be thought of as sampling from an infinite target population in which each unit appears an inexhaustible number of times. Now sampling from an infinite population implies that the finite population correction $(1 - n/N)$ is close to one. Furthermore, when N is very large, N is approximately equal to $N - 1$ and as a result S^2 is approximately equal to σ^2. Altogether, the formula in Expression 10.2 reduces to the formula in Expression 10.1.

Before concluding this introduction to sampling with replacement, we have three remarks. First, when sampling with replacement, it is possible for the sample size to be greater than the population size and still not have observed *every* unit in the finite population! In the taxpayer example, a sample 8, 9 or even 10 of the 8 taxpayers need not include all 8 of them. This cannot occur when sampling without replacement. In that case all 8 taxpayers must be included.

Second, the variance of the mean when sampling with replacement is always larger than when sampling without replacement. This is shown in Section 10.1.4.

Finally, in the more mathematical presentations of statistical methods, discussion of sampling with replacement is much more common than sampling without replacement. There are at least two reasons for this. One is that it is popular to assume that the sampled population is infinite in size. The other is that sampling with replacement is associated with more tractable mathematics.

10.1.3 Relative Frequencies

Examination of Table 10.1 shows that some of the *ordered* pairs of taxpayers give rise to the same *unordered* sample pair and as a result the same sample mean. For example, the sample of taxpayer 1 and taxpayer 8 can arise by the selection of taxpayer 1 first or by taxpayer 8 first. In both cases, a sample mean of 95 is obtained. This unordered sample is listed in two places in Table 10.1, once in the first set of eight ordered pairs and once in the last set. On the other hand, samples with the same taxpayer repeated twice appear only once among the 64 ordered pairs. Clearly, some sample means are more likely to appear than others. This feature can be seen in Figure 10.1, in which the means appear with different frequencies.

The duplication of some samples permits a simplification in the calculation of $E(\bar{y})$ and $E(\bar{y}-\bar{Y})^2$. It will turn out that this simplification is also very informative. First consider the following calculations.

$$E(\bar{y}) = \frac{1}{64}\,[60.0 + 2(66.0) + 2(64.0) + \cdots + 130.0]$$

$$= \frac{1}{64}\,(60.0) + \frac{2}{64}\,(66.0) + \frac{2}{64}\,(64.0) + \cdots + \frac{1}{64}\,(130.0)$$

$$= 91.50 \tag{10.3}$$

and

$$E(\bar{y} - \bar{Y})^2 = \frac{1}{64}\,[(60.0-91.5)^2 + 2(66.0-91.5)^2$$

$$+ 2(64.0-91.5)^2 + \cdots + (130.0-91.5)^2]$$

$$= \frac{1}{64}\,(60.0-91.5)^2 + \frac{2}{64}\,(66.0-91.5)^2$$

$$+ \frac{2}{64}\,(64.0-91.5)^2 + \cdots + \frac{1}{64}\,(130.0-91.5)^2$$

$$= 257.88. \tag{10.4}$$

In the preceding first line of $E(\bar{y})$, those sample means that appear twice in the sampling distribution are written down only once but with a weight of 2. In the next line, the divisor of 64 (the number of equally likely sample means) is taken inside the square brackets and the calculation is equivalently written by weighting the possible sample values either by 1/64 or 2/64. These weights are the *relative frequencies* with which the particular sample means occur. A similar discussion applies to the calculation of $E(\bar{y}-\bar{Y})^2$. This mode of calculation of $E(\bar{y})$ and $E(\bar{y}-\bar{Y})^2$ will be very useful to us later.

The compression of the calculations described in Expressions 10.3 and 10.4 can also be used to compress Table 10.1 into Table 10.2. Table 10.2 is another tabular representation of the with replacement sampling distribution of $\bar{y}$. A plot of this sampling distribution is, of course, the same as the one given in Figure 10.1.

It is important at this point to make a certain comparison of the *with* replacement sampling distribution of $\bar{y}$, Table 10.2, and the *without* replacement sampling distribution of $\bar{y}$, Table 5.2. First notice that there are 36 different samples listed in Table 10.2; these are the

TABLE 10.2

WITH REPLACEMENT SAMPLING DISTRIBUTION OF $\bar{y}$
[2 taxpayers from 8]

Sample Number	Sampled Taxpayers	Sample Mean	Frequency	Relative Frequency
1	1,2	66.0	2	2/64
2	1,3	64.0	2	2/64
3	1,4	77.0	2	2/64
4	1,5	75.0	2	2/64
5	1,6	81.0	2	2/64
6	1,7	88.0	2	2/64
7	1,8	95.0	2	2/64
8	2,3	70.0	2	2/64
9	2,4	83.0	2	2/64
10	2,5	81.0	2	2/64
11	2,6	87.0	2	2/64
12	2,7	94.0	2	2/64
13	2,8	101.0	2	2/64
14	3,4	81.0	2	2/64
15	3,5	79.0	2	2/64
16	3,6	85.0	2	2/64
17	3,7	92.0	2	2/64
18	3,8	99.0	2	2/64
19	4,5	2.0	2	2/64
20	4,6	98.0	2	2/64
21	4,7	105.0	2	2/64
22	4,8	112.0	2	2/64
23	5,6	96.0	2	2/64
24	5,7	103.0	2	2/64
25	5,8	110.0	2	2/64
26	6,7	109.0	2	2/64
27	6,8	116.0	2	2/64
28	7,8	123.0	2	2/64
29	1,1	68.0	1	1/64
30	2,2	72.0	1	1/64
31	3,3	68.0	1	1/64
32	4,4	90.0	1	1/64
33	5,5	90.0	1	1/64
34	6,6	102.0	1	1/64
35	7,7	116.0	1	1/64
36	8,8	130.0	1	1/64
TOTALS		5856.0	64	1

$$E(\bar{y}) = 91.5 = \bar{Y}$$

$$E(\bar{y} - \bar{Y})^2 = 257.88 = 515.75/2 = \sigma^2/n$$

$\binom{8}{2} = 28$ samples with no repetitions, that is, the samples that are possible when sampling without replacement, *plus* the eight samples in which the units are repeated. Also, in Table 5.2, each of the possible sample means is as likely to be selected as any other. But in Table 10.2 the samples without repetitions are twice as likely to occur as the samples with repetitions. In later sections we encounter more complicated sampling distributions with this same characteristic that different samples occur with different frequencies.

10.1.4 In Practice

Algebraically, it is easy to see that the ratio of the variance of the sample mean when sampling without replacement, $(1 - n/N)S^2/n$, to the variance with replacement σ^2/n, is equal to $(N-n)/(N-1)$. Since $N - n$ is less than $N - 1$ for $n > 1$, this ratio of the variances is less than one when $n > 1$. As a result, sampling without replacement almost always has a smaller variance than sampling with replacement. The only exception is the trivial case when $n = 1$. Furthermore, for N even moderately large, the ratio of the variances is very close to the finite population correction, $(N-n)/N$, (5.4). Consequently, we can think of the gain due to sampling without replacement as the effect of the finite population correction on the variance of the estimator. When the population size is large relative to the sample size, the finite population correction is close to one and the variance formulas for with and without replacement are effectively the same. Another way of looking at this is that in large populations from which a relatively small sample is drawn, the difference between sampling with and without replacement is small because samples with repetitions have a very small probability of occurring.

On the other hand, the difference between sampling with and without replacement is significant if n/N is large. This happens if the population is small and/or the sample large. In these cases, a general mathematical result tells us that if repetitions occur, then it pays to ignore them. For example, if a sample of size 8 is selected with 7 distinct units one of which is repeated to complete the sample of 8, it is more efficient to ignore the repetition and treat the sample as a sample of size 7 *without* replacement. Even though this may seem counterintuitive, it can be shown that such a sampling procedure leads to a variance that is smaller, or at least not larger, than the variance of the estimator that includes the repetitions. We are not going to pursue this point further in this book, but it is a matter for investigation should the situation be encountered in practice.

In this section, it was shown that changing the method of sampling changes the sampling distribution of $\bar{y}$. This includes the variance of $\bar{y}$. Unfortunately, sampling with replacement, the sampling scheme newly introduced in this chapter, has a larger variance than sampling without replacement. But now an important question can be raised, namely can methods of sampling be found which have unbiased estimators (or almost so) *and* smaller sampling variances? We shall see that they can.

Exercises

10.1.1   List in systematic manner all the possible samples of size 3 that can be selected by sampling with equal probability *with* replacement from the following small population of size 5,

$$y_1 = 6, \ y_2 = 2, \ y_3 = 5, \ y_4 = 12, \ y_5 = 10.$$

(This is the same population that was considered in various exercises in Chapter 5 beginning with Exercise 5.2.1.)

10.1.2   Compute the sample mean, $\bar{y}$, and the sample median, $\hat{y}_{med}$, for each of the 125 samples listed in Exercise 10.1.1.

10.1.3   Calculate both $E(\bar{y})$ and $E(\hat{y}_{med})$ directly from the sampling distributions of Exercise 10.1.2.

10.1.4   Calculate both $Var(\bar{y})$ and $Var(\hat{y}_{med})$ directly from the sampling distribution of Exercise 10.1.2.

10.1.5   Calculate $Var(\bar{y})$ for the population in Exercise 10.1.1 directly by the formula given in Equation 10.1. Compare this calculation with the result obtained for $Var(\bar{y})$ in Exercise 10.1.4.

10.1.6   Compare $Var(\bar{y})$ for sampling with replacement (Exercises 10.1.4 and 10.1.5) with $Var(\bar{y})$ for sampling without replacement (Exercise 5.4.1). What is the ratio of the "with replacement" variance to the "without replacement" variance?

10.1.7   Referring to Exercise 10.1.6, if the population were of size $N=500$, that is, 100 times as large, yet with the same values of $\bar{Y}$, S^2, and σ^2, calculate

   i.   the "with replacement" variance of $\bar{y}$

    ii.   the "without replacement" variance of $\bar{y}$

    iii. the ratio of the variance in (i) to variance in (ii). Explain why this ratio is different from the ratio calculated in Exercise 10.1.6.

10.1.8   Go through the sampling distribution in Exercise 10.1.1 and eliminate all repeated observations. For example, the sample (y_1, y_1, y_3) of size $n=3$ becomes the sample (y_1, y_3) of size $n=2$. Then for each of the samples that have "changed" recalculate a new sample mean based only on the unique observations and not on the repeated observations.

10.1.9   Calculate $E(\bar{y})$ and $Var(\bar{y})$ for the sampling distribution in Exercise 10.1.8. How do these values compare with those obtained in Exercise 10.1.4 for sampling with replacement and Exercise 5.4.1 for sampling without replacement? Discuss your comparisons.

10.1.10 In Exercise 3.2.1, it was pointed out that taxpayers T_1, T_2, T_3, T_5, and T_9 are Democrats and the rest Republicans. Use the results of Exercises 3.2.1 and 3.2.5 to show that the sampling variance of p is given by

$$Var(p) = (1 - \frac{n}{N}) \frac{1}{n} \frac{NPQ}{N-1} \, ,$$

where p and $P = 1 - Q$ are the sample and population proportions of Democrats, respectively. Sampling is without replacement. Then show that the sample variance of p is

$$var(p) = (1 - \frac{n}{N}) \frac{pq}{n-1} \, .$$

10.1.11 A true die has one of the "numbers" 1, 2, 3, 4, 5, 6 on each of its 6 sides. No "number" appears twice. On a random toss of the die each of these numbers is equally likely to appear face up. What is the expected value of a single random roll?

10.1.12 A die was accidentally made incorrectly. The "3" that was supposed to be on one face actually appeared as a "5."
(a) What is the expected result of a single toss of this die?
(b) What is the bias in this faulty die when compared with an accurately made die?

10.1.13 Calculate the variance of the expected values described in Exercises 10.1.11 and 10.1.12.

10.1.14 A sample can be drawn from a finite population by sampling
with replacement or sampling without replacement. If a sample of size 15 is drawn without replacement from a population
of size 60, the sampling variance of the mean has a certain
value. How much larger must the sample size be when sampling *with* replacement in order that the variance of the mean
be of this same value?

10.2 Sampling with Unequal Probability

In Section 10.1, sampling with equal probability and with replacement was discussed. Unfortunately, we saw that the variance of the
mean is larger when sampling with replacement than it is when sampling without replacement. However, some knowledge of sampling
with equal probabilities and with replacement makes it easier to discuss
sampling with *unequal probabilities,* and this type of sampling can be.
much better than any that we have discussed so far. Sampling with
unequal probability is discussed in this section.

10.2.1 A Simple Example

In Section 10.1 the sampling procedures were implemented by
placing eight discs, *one* for each of the taxpayers, face down on a table.
What happens to the characteristics of the sampling process if *two* discs
are placed on the table representing taxpayer number 1? And in general, what happens if arbitrary and unequal numbers of discs representing each taxpayer are placed on the table?

First, consider the simple case of a single extra disc representing
taxpayer 1. It will turn out that this is an undesirable thing to do, but
discussing this case is very enlightening. The sampling procedure
specifies the selection of two discs randomly with replacement from the
nine on the table. Keep in mind that the selection of disc nine implies
the selection of taxpayer 1. As a result, on *both* the first and second
draws taxpayer 1 has twice the chance of selection (2/9 to 1/9) of the
other seven taxpayers.

Since there are nine discs, each of which is equally likely to be
selected, there are 9×9 equally likely ordered pairs of taxpayers that can
be selected. All 81 ordered pairs appear in Table 10.3. Table 10.3 is
exactly analogous to Table 10.1 in which the 8×8 pairs arising from

TABLE 10.3

UNEQUAL CHANCE SELECTION OF 2 TAXPAYERS FROM 8
[81 equally likely ordered pairs]
[Taxpayer listed first is selected first]

Taxpayers Selected	Sample Mean	Taxpayers Selected	Sample Mean	Taxpayers Selected	Sample Mean	Taxpayers Selected	Sample Mean
1,1	60.0	2,1	66.0	3,1	64.0	4,1	77.0
1,2	66.0	2,2	72.0	3,2	70.0	4,2	83.0
1,3	64.0	2,3	70.0	3,3	68.0	4,3	81.0
1,4	77.0	2,4	83.0	3,4	81.0	4,4	94.0
1,5	75.0	2,5	81.0	3,5	79.0	4,5	92.0
1,6	81.0	2,6	87.0	3,6	85.0	4,6	98.0
1,7	88.0	2,7	94.0	3,7	92.0	4,7	105.0
1,8	95.0	2,8	101.0	3,8	99.0	4,8	112.0
1,9(1)	60.0	2,9(1)	66.0	3,9(1)	64.0	4,9(1)	77.0
5,1	75.0	6,1	81.0	7,1	88.0	8,1	95.0
5,2	81.0	6,2	87.0	7,2	94.0	8,2	101.0
5,3	79.0	6,3	85.0	7,3	92.0	8,3	99.0
5,4	92.0	6,4	98.0	7,4	105.0	8,4	112.0
5,5	90.0	6,5	96.0	7,5	103.0	8,5	110.0
5,6	96.0	6,6	102.0	7,6	109.0	8,6	116.0
5,7	102.0	6,7	109.0	7,7	116.0	8,7	123.0
5,8	110.0	6,8	116.0	7,8	123.0	8,8	130.0
5,9(1)	75.0	6,9(1)	81.0	7,9(1)	88.0	8,9(1)	95.0

Taxpayers Selected	Sample Mean
9(1),1	60.0
9(1),2	66.0
9(1),3	64.0
9(1),4	77.0
9(1),5	75.0
9(1),6	81.0
9(1),7	88.0
9(1),8	95.0
9(1),9(1)	60.0

$$E(\bar{y}) = 88.0$$
$$E(\bar{y} - \bar{Y})^2 = 278.22$$

TABLE 10.4

UNEQUAL CHANCE SELECTION OF 2 OF 8 TAXPAYERS

Sample Number	Sample Units	Sample Mean $\bar{y}$	Weighted Mean $\bar{y}_p$	Frequency	Relative Frequency
1	1,2	66.0	57.4	4	4/81
2	1,3	64.0	55.1	4	4/81
3	1,4	77.0	69.8	4	4/81
4	1,5	75.0	67.5	4	4/81
5	1,6	81.0	74.3	4	4/81
6	1,7	88.0	82.1	4	4/81
7	1,8	95.0	90.0	4	4/81
8	2,3	70.0	78.8	2	2/81
9	2,4	83.0	93.4	2	2/81
10	2,5	81.0	91.1	2	2/81
11	2,6	87.0	97.9	2	2/81
12	2,7	94.0	105.8	2	2/81
13	2,8	101.0	113.6	2	2/81
14	3,4	81.0	91.1	2	2/81
15	3,5	79.0	88.9	2	2/81
16	3,6	85.0	95.6	2	2/81
17	3,7	92.0	103.5	2	2/81
18	3,8	99.0	111.4	2	2/81
19	4,5	92.0	103.5	2	2/81
20	4,6	98.0	110.3	2	2/81
21	4,7	105.0	118.1	2	2/81
22	4,8	112.0	126.0	2	2/81
23	5,6	96.0	108.0	2	2/81
24	5,7	103.0	115.9	2	2/81
25	5,8	110.0	123.8	2	2/81
26	6,7	109.0	122.6	2	2/81
27	6,8	116.0	130.5	2	2/81
28	7,8	123.0	138.4	2	2/81
29	1,1	60.0	33.8	4	4/81
30	2,2	72.0	81.0	1	1/81
31	3,3	68.0	76.5	1	1/81
32	4,4	94.0	105.8	1	1/81
33	5,5	90.0	101.3	1	1/81
34	6,6	102.0	114.8	1	1/81
35	7,7	116.0	130.5	1	1/81
36	8,8	130.0	146.3	1	1/81

TOTALS   81   1

$$E(\bar{y}) = 88.0 \qquad E(\bar{y}_p) = 91.5$$

$$E(\bar{y} - \bar{Y})^2 = 278.22 \qquad E(\bar{y}_p - \bar{Y})^2 = 686.81$$

eight discs were displayed. A difference is that in Table 10.3, when disc 9 appears, a (1) is placed along side the "9" as a reminder that disc 9 implies the selection of taxpayer 1.

The values of $\bar{y}$ in Table 10.3 constitute all the possible values of $\bar{y}$ that can occur, that is, the sampling distribution of $\bar{y}$.

In Section 10.1, Table 10.1 was compressed into Table 10.2 by using the knowledge that the sample mean is independent of the order of selection of the taxpayers. For example, if taxpayer 2 is drawn first followed by taxpayer 3, the same sample mean is obtained by their selection in the reverse order. This compression of Table 10.1 made it clear that samples with distinct units appeared with a 2/64 chance and samples with repeated units appeared with a 1/64 chance.

The inclusion of two discs representing taxpayer 1 has changed this particular symmetry. The sample mean $\bar{y} = 60.0$ calculated from the sample $(y_1 = 60,\ y_1 = 60)$ now has not 1, but 4, separate, equal chances of occurring. The other sample means are also observed with different frequencies. By directly counting the occurrences in Table 10.3, the frequencies tabulated in Table 10.4 can be verified. Table 10.4 displays the same sampling distribution as Table 10.3 but in a more compressed form. Both tables detail the sampling distribution of $\bar{y}$ when taxpayer 1 has a 2/9 chance of selection and all others 1/9.

In Table 10.4 the frequency with which each sample mean occurs is obtained directly from Table 10.3 by counting. The relative frequencies are calculated by dividing the counted frequencies by the total possible number of equally likely samples, in this case 81. So, for example, the relative frequency of the sample mean $\bar{y} = 66.0$ is 4/81 and the relative frequency of $\bar{y} = 85.0$ is 2/81.

It turns out that it is not necessary to obtain the *relative* frequencies by first counting the equally likely samples. The relative frequencies can be obtained directly by multiplication of the *individual* chances of selection. To illustrate, the chance of getting taxpayer 2 on *both* the first and second draw is $1/9 \times 1/9 = 1/81$, the same result obtained in Table 10.4 by direct counting. Samples without repetitions are slightly more complicated. The chance of selecting taxpayer 1 on the first draw is 2/9 and taxpayer 2 on the second is 1/9, so the combined chance of getting *both* taxpayer 1 on the first draw *and* taxpayer 2 on the second is $2/9 \times 1/9 = 2/81$. But in addition, there is the same chance $(1/9 \times 2/9)$ of getting these two taxpayers in the *reverse* order, that is, with taxpayer 2 first. So the chance of getting $\bar{y} = 66$ is $2/81 + 2/81 = 4/81$, which again checks with Table 10.4. By the same reasoning, the chance of getting $\bar{y} = 85.0$ from taxpayers 3 and 6 is $1/81 + 1/81 = 2/81$.

As we shall see in a later example, the ability to determine these relative frequencies by multiplication is very important in practice because direct counting becomes a virtual impossibility.

The next important feature to notice about the sampling distribution in Tables 10.3 and 10.4 is that $E(\bar{y}) = 88.00 \neq 91.50$. Consequently, the sample mean $\bar{y}$ is now a biased estimator of $\bar{Y}$. The average value of $\bar{y}$ is 3.50 lower than $\bar{Y}$. This underestimation is a direct result of T_1, with his low value of $y_1 = 60.0$, appearing more often than the others in the sampling distribution. As a result, $E(\bar{y})$ is pulled down below the true mean of 91.5 to the lower value of 88.0.

To better understand the bias caused by the over-representation of taxpayer 1, recall that the *equal* chance sampling of 2 of 8 taxpayers leads to $\bar{y}$ being an unbiased estimator of $\bar{Y}$. Moreover, this is a general result, the equal chance selection of a sample of n from a population of N makes $\bar{y}$ an unbiased estimator. Consequently, by analogy in the current example, and since the 9 *discs* are selected with equal chances, we know that $\bar{y}$ is an unbiased estimator of the mean of the 9 *numbers associated with the 9 discs*. The value $y_1 = 60$ appears twice, once in association with the first disc and once in association with the ninth. So, $\bar{y}$ is an unbiased estimator as follows.

$$E(\bar{y}) = \frac{1}{9}[60+72+68+94+90+102+116+130+60]$$

$$= 88.0. \tag{10.5}$$

This calculation agrees numerically, as it must, with the earlier calculation (Table 10.3 or 10.4) of $E(\bar{y})$ by averaging over all the possible equally likely sample means.

Now rewriting Equation 10.5 as,

$$E(\bar{y}) = \frac{1}{9}[2(60)+72+68+94+90+102+116+130]$$

$$= \frac{2}{9}(60) + \frac{1}{9}(72) + \frac{1}{9}(68) + \frac{1}{9}(94)$$

$$+ \frac{1}{9}(90) + \frac{1}{9}(102) + \frac{1}{9}(116) + \frac{1}{9}(130)$$

$$= 88.0, \tag{10.6}$$

shows that taxpayer 1 is weighted too heavily in $E(\bar{y})$ by an amount proportional to its *increased chance of selection,* 2/9 to 1/9 or equivalently 2 to 1. This observation suggests that a suitable correction for the bias brought about by permitting taxpayer 1 to enter the sample more frequently than the others might be obtained by *weighting* taxpayer 1 inversely to his chance of selection. Specifically, this suggestion would require weighting taxpayer 1 *half* as much as the other taxpayers whenever taxpayer 1 appears in the sample. As we shall see shortly, such a weighting is exactly the correct procedure to obtain an unbiased estimator.

At this point, it is useful to have some general notation. If the chance of selecting taxpayer i is p_i, then the idea of weighting each taxpayer *inversely* to his chance of selection suggests using an estimate proportional to

$$\frac{1}{2} \left(\frac{y_i}{p_i} + \frac{y_j}{p_j} \right) \tag{10.7}$$

instead of the equally weighted mean,

$$\bar{y} = \frac{y_i + y_j}{2}. \tag{10.8}$$

A small but important point about Expression 10.7 is that it estimates the total Y and *not* the mean $\bar{Y}$! To see that this is so, set all $p_i = 1/8$, which is the equal chance case discussed in Section 10.1, then Expression 10.7 is equal to

$$\frac{1}{2} \left(\frac{y_i}{1/8} + \frac{y_j}{1/8} \right) \tag{10.9}$$

$$= 8\bar{y} = N\bar{y} \ .$$

This makes it clear that a division by N is necessary if the true *mean* is to be estimated rather than the true *total*. Therefore, it is the estimator,

$$\bar{y}_p = \frac{1}{8} \frac{1}{2} \left(\frac{y_i}{p_i} + \frac{y_j}{p_j} \right), \tag{10.10}$$

which results from the suggestion that reciprocal weighting might create an unbiased estimator of the mean, $\bar{Y}$.

To illustrate, if taxpayers 1 and 7 are selected, the weighted estimate is

$$\bar{y}_p = \frac{1}{8} \frac{1}{2} \left(\frac{60}{2/9} + \frac{116}{1/9} \right) = 82 \ 1/8$$

which can be compared with the equally weighted mean

$$\bar{y} = \frac{1}{2} \ (60+116) = 88.$$

As anticipated, the unequal weighting of taxpayer 1 makes the estimator $\bar{y}_p$ unbiased. This can be confirmed numerically in Table 10.4 and can also be proven mathematically. So the general conclusion is that, if one population unit has a higher chance of entering the sample, an unbiased estimator can be obtained by giving that unit a weight in the estimation formula which is inversely proportional to its chance of selection. Such a weighting compensates for over-representation of that unit in the sample draw.

Now that we have constructed $\bar{y}_p$ as an unbiased estimator of $\bar{Y}$, we might ask why, in the preceding calculation of $\bar{y}_p$ and $\bar{y}$ for taxpayers 1 and 7, is $\bar{y}_p$ farther away from the true mean, 91.5, than the equally weighted mean $\bar{y}$? This is an important question. To answer it, we need to consider the variance of $\bar{y}_p$. But as we warned at the outset of this section, giving taxpayer 1 twice the chance of entering the sample is not a desirable thing to do.

The variance of $\bar{y}_p$, is given in Table 10.4. It is calculated in the usual way as,

$$E(\bar{y}_p - \bar{Y})^2 = \frac{4}{81}(57.4 - 91.5)^2 + \frac{4}{81}(55.1 - 91.5)^2 + \cdots + \frac{1}{81}(146.3 - 91.5)^2$$

$$= 686.61 \tag{10.11}$$

This result comes from direct calculation in either Table 10.3 or 10.4. But as usual, there is also a formula which can be used to give the *same* result. In this case, statisticians have proven that

$$E(\bar{y}_p - \bar{Y})^2 = \frac{1}{N^2}\frac{1}{n}\sum_{i=1}^{N}p_i\left(\frac{y_i}{p_i} - Y\right)^2$$

$$= \frac{1}{8^2}\frac{1}{2}\left\{\frac{2}{9}\left(\frac{60}{2/9} - 732\right)^2 + \cdots + \frac{1}{9}\left(\frac{130}{1/9} - 732\right)^2\right\}$$

$$= 686.61. \tag{10.12}$$

We shall discuss the variance of $\bar{y}_p$ extensively in the next section.

10.2.2 Sampling with Probabilities Proportional to Size

It is now apparent that an unfortunate thing has occurred. While we were busy creating an unbiased estimator to account for the over-representation of taxpayer 1, we also created an estimator with a larger variance. The variance of $\bar{y}$ when sampling with equal probability is 257.88 (see Expression 10.4), but when T_1 has twice the chance of the other taxpayers of entering the sample, the variance of $\bar{y}_p$ is 686.81 (see Expression 10.11 or 10.12). Regrettably, a new source of variability has been added to the estimator. It is the probabilities of selection, the p_i's, so that there are now two sources of variability, the y_i's *and* the p_i's. To see that the p_i's are indeed a source of variability, imagine a population in which all eight taxpayers have the *same* income, say $y_i = 1$. In such a population, a reasonable estimator would have zero sampling variance for the simple reason that the population has zero variance. But unfortunately, $\bar{y}_p$ does not have zero variance in this case. To confirm this, consider the two samples of taxpayers 1 and 3, and taxpayers 2 and 3. For the first sample

$$\bar{y}_p = \left(\frac{1}{8}\right) \left(\frac{1}{2}\right) \left(\frac{1}{2/9} + \frac{1}{1/9}\right) = \frac{27}{32},$$

and for the second

$$\bar{y}_p = \frac{1}{8} \frac{1}{2} \left(\frac{1}{1/9} + \frac{1}{1/9}\right) = \frac{36}{32} = \frac{18}{16}.$$

Since these estimates are not equal, it means that $\bar{y}_p$ does not have a zero sampling variance even when the sampled population contains only identical units!

It is worth remarking at this point that a comparison of the biased estimator $\bar{y}$ with the unbiased estimator $\bar{y}_p$ on the basis of mean square error would favor the biased estimator $\bar{y}$! Simple calculations show that

$$MSE(\bar{y}) = Var(\bar{y}) + [Bias(\bar{y})]^2$$
$$= 278.2 + [88.0 - 91.5]^2$$
$$= 290.5 \tag{10.13}$$

and

$$MSE(\bar{y}_p) = Var(\bar{y}_p) + [Bias(\bar{y}_p)]^2$$
$$= 686.8 + [91.5 - 91.5]^2$$
$$= 686.8. \tag{10.14}$$

Can anything be done about the large variance in $\bar{y}_p$? More specifically, can the unbiasedness of the estimator be retained without this unfortunate increase in variance over $\bar{y}$? Or better yet, can the variance be reduced? The answer is a very definite yes.

To see how to reduce variance, suppose that the chances of selection are proportional to the measurement, $p_i = y_i/Y$. In the taxpayer example, $Y = 732$, so,

$$p_1 = \frac{60}{732}, \ p_2 = \frac{72}{732}, \ p_3 = \frac{68}{732}, \ p_4 = \frac{94}{732},$$

$$p_5 = \frac{90}{732}, \ p_6 = \frac{102}{732}, \ p_7 = \frac{116}{732}, \ p_8 = \frac{130}{732}. \tag{10.15}$$

To do this operationally in the same way as the earlier examples would require placing 732 discs on the table for selection, 60 of which represent the first taxpayer, 72 of which represent the second taxpayer, and so on for all eight taxpayers. However, in practice, it is easier to use one of the many sets of published random integer tables, letting the selection of integers from 1 to 60 represent taxpayer 1, from 61 to $60 + 72 = 132$ taxpayer 2, and so on as required.

But no matter how the sample selection is operationally implemented, if values of p_i are used which are proportional to y_i a startling thing happens, the unbiased estimator achieves a *zero* variance! To see this, suppose that taxpayers 1 and 2 are selected. Then,

$$\bar{y}_p = \frac{1}{8} \frac{1}{2} \left(\frac{60}{60/732} + \frac{72}{72/732} \right)$$

$$= \frac{1}{8} \frac{1}{2} \ (732{+}732) = \frac{732}{8} = 91.5,$$

which is *exactly* equal to the true mean. If taxpayers 4 and 5 are selected,

$$\bar{y}_p = \frac{1}{8} \frac{1}{2} \left(\frac{94}{94/732} + \frac{90}{90/732} \right) = 91.5,$$

which is again exactly equal to the true mean! Furthermore, this characteristic is no accident; whichever sample is selected, $\bar{y}_p$ is always equal to 91.50. So in this case, $\bar{y}_p$ is exactly unbiased and has zero variance as well! What has happened is that the new source of variability, the p_i's, has been used to literally cancel out the variability in the y_i's.

But regrettably, this zero variance estimator is too good to be true, and the reason is obvious. In practice, the y_i values are *not* known in advance of sampling, and consequently, cannot be used to draw the sample. The prospect of a zero variance estimator was indeed too much to expect. But all is not lost nor is it anywhere near being all lost, for while the y_i values are not known in advance of sampling, measurements that are highly similar to the y_i are often available. As one might hope from the preceding zero variance case of $\bar{y}_p$, the selection of sample units with probabilities proportional to available, *similar* measurements can reduce variance substantially.

In the taxpayer example, X_i, the reported income, may reasonably be expected to be very similar to y_i, the actual income. Hence, we can consider drawing the sample with chances proportional to x_i rather than y_i, hoping that this will result in most of the reduction in variance that we saw was possible by using y_i. In the taxpayer example $X = 660$, so the probabilities which are proportional to x_i are,

$$p_1 = \frac{50}{660} \quad p_2 = \frac{56}{660} \quad p_3 = \frac{66}{660} \quad p_4 = \frac{76}{660}$$

$$p_5 = \frac{90}{660} \quad p_6 = \frac{100}{660} \quad p_7 = \frac{112}{660} \quad p_8 = \frac{110}{660},$$

(10.16)

where each $p_i = x_i/X$ (see Table 5.1). So if taxpayers 1 and 2 are selected, then

$$\bar{y}_x = \frac{1}{8}\frac{1}{2}(\frac{60}{50/660} + \frac{72}{56/660}) = 102.54,$$

and if taxpayers 4 and 5 are selected,

$$\bar{y}_x = \frac{1}{8}\frac{1}{2}(\frac{94}{76/660} + \frac{90}{90/660}) = 92.27.$$

The entire sampling distribution $\bar{y}_p$ $(p_i = x_i/X)$ is included in Table 10.5. To distinguish this special case, the estimator is labeled as $\bar{y}_x$.

Now let us go back to the sample of taxpayers 1 and 7 which we discussed in Section 10.2.1. For this sample,

$$\bar{y}_x = \frac{1}{8}\frac{1}{2}(\frac{60}{50/660} + \frac{116}{112/660}) = 92.22.$$

Recall that when taxpayer 1 was sampled with probability 2/9 and all the others with probability 1/9, an unbiased estimator was achieved but the weighting also took $\bar{y}_p$ *farther* away from the true mean 91.5 than the equally weighted mean. Specifically, the weighted estimator was 82 1/8 and the equally weighted mean was 88.0. This resulted from the large variance that $\bar{y}_p$ had in the example of Section 10.2.1. In the example of this section, weighting by probabilities proportional to x_i/X does not increase variance in this unfortunate way. In fact, as we shall see, the variance is decreased enormously.

TABLE 10.5

SAMPLING DISTRIBUTION OF $\bar{y}_x$

(Chance of Selection Proportional to Reported Income, x)

Sample Number	Sampled Taxpayers	$\bar{y}_x$	Relative Frequency	
1	1,2	102.5	2(50/660)(56/660)	= 0.013
2	1,3	92.0	2(50/660)(66/660)	= 0.015
3	1,4	100.5	2(50/660)(76/660)	= 0.017
4	1,5	90.8	2(50/660)(90/660)	= 0.021
5	1,6	91.6	2(50/660)(100/660)	= 0.023
6	1,7	92.2	2(50/660)(112/660)	= 0.026
7	1,8	98.3	2(50/660)(110/660)	= 0.025
8	2,3	95.5	2(56/660)(66/660)	= 0.017
9	2,4	104.1	2(56/660)(76/660)	= 0.020
10	2,5	94.3	2(56/660)(90/660)	= 0.023
11	2,6	95.1	2(56/660)(100/660)	= 0.026
12	2,7	95.8	2(56/660)(112/660)	= 0.029
13	2,8	101.8	2(56/660)(110/660)	= 0.028
14	3,4	93.5	2(66/660)(76/660)	= 0.023
15	3,5	83.8	2(66/660)(90/660)	= 0.027
16	3,6	84.6	2(66/660)(100/660)	= 0.030
17	3,7	85.2	2(66/660)(112/660)	= 0.034
18	3,8	91.3	2(66/660)(110/660)	= 0.033
19	4,5	92.3	2(76/660)(90/660)	= 0.031
20	4,6	93.1	2(76/660)(100/660)	= 0.035
21	4,7	93.7	2(76/660)(112/660)	= 0.039
22	4,8	99.8	2(76/660)(110/660)	= 0.038
23	5,6	83.3	2(90/660)(100/660)	= 0.041
24	5,7	84.0	2(90/660)(112/660)	= 0.046
25	5,8	90.0	2(90/660)(110/660)	= 0.045
26	6,7	84.8	2(100/660)(112/660)	= 0.051
27	6,8	90.8	2(100/660)(110/660)	= 0.051
28	7,8	91.5	2(112/660)(110/660)	= 0.057
29	1,1	99.0	$(50/660)^2$	= 0.006
30	2,2	106.1	$(56/660)^2$	= 0.007
31	3,3	85.0	$(66/660)^2$	= 0.010
32	4,4	102.0	$(76/660)^2$	= 0.013
33	5,5	82.5	$(90/660)^2$	= 0.019
34	6,6	84.1	$(100/600)^2$	= 0.023
35	7,7	85.4	$(112/660)^2$	= 0.029
36	8,8	97.5	$(110/660)^2$	= 0.028

TOTAL 1

$$E(\bar{y}_x) = 91.5$$

$$Var(\bar{y}_x) = E(\bar{y}_x - \bar{Y})^2 = 35.37$$

Table 10.5 also contains the relative frequencies with which each of the samples appears. In our earlier simple example, these relative frequencies were obtained by first actually enumerating the samples. In the current example, this would require counting and classifying $660 \times 660 = 435,600$ cases which is not impossible, but in real applications is hopelessly cumbersome. Fortunately, as we also saw earlier, enumeration is quite unnecessary because the required frequencies can be calculated by multiplying together the individual chances of selection. For example, the selection of taxpayer 1 followed by taxpayer 2 occurs with a relative frequency of $(50/660)(56/660)$. The same relative frequency refers to the units drawn in the reverse order so that $\bar{y}_x = 102.5$ appears with a relative frequency of $2(50/660)(56/660)$.

As expected, the weighting of each taxpayer inversely to his chance of selection ensures that $\bar{y}_x$ is unbiased; see Table 10.5. In addition, the variance of $\bar{y}_x$ is 35.37 which is much smaller than almost any scheme we have considered so far. The exceptional scheme had a variance equal to zero, but was not practically useful. So it has turned out that the known reported incomes (x_i) are an excellent surrogate for the unknown actual incomes, (y_i), and have been used in $\bar{y}_x$ to reduce the sampling variance substantially.

In summary, sampling with unequal probabilities can be used to make dramatic reductions in variance. This is a major lesson of sample design. It is important to remember however, that this new tool can be dangerous. First, if the chances of selection are not proportional to the actual measurements, the technique can *increase* variance. This is exactly what happened when taxpayer 1 (who has the smallest income) was given twice the chance of the other taxpayers of entering the sample. Second, care must be taken to weight the observations inversely to their chance of selection. Otherwise the estimators will be biased.

It is very important to understand sampling with probabilities proportional to size. It has important practical implications.

Exercises

10.2.1   Show that the formula for $Var(\hat{y}_p)$ is equal to zero when the probabilities of selection are proportional to size, that is $p_i = y_i/Y$.

10.2.2   Samples of farms are sometimes drawn by randomly selecting points on an area map. If a selected point falls within the map boundaries of a farm, then that farm is included in the sample. Suppose that seven points are selected in this way and the following sample information is obtained:

Point Number	Farm Acreage	Acreage in Tobacco
1	100	70
2	110	51
3	1300	150
4	97	50
5	48	24
6	(Same farm as point number 4)	
7	29	0

(a) Utilize the preceding information to estimate the average acreage per farm in tobacco.
(b) Estimate the variance of the estimate in (a).

10.3 Stacking the Deck

10.3.1 Selection Biases

Intuitively, everyone knows that "stacking the deck" means manipulating the outcome of a game or race in such a way that the odds of winning are tipped more in one direction than another. The usual places are card and dice games, but "stacking the deck" occurs in applied statistics too.

To make the discussion concrete, if I take an even bet that a single card *randomly* picked from a deck is red and then, before the selection, proceed to remove half a dozen black cards, I have stacked the deck in my favor. The chances in the fair bet (that you *think* you have) are $26/52 = 1/2$ for red and $26/52 = 1/2$ for black, but the chances are now really $26/46 = 13/23$ for red and $20/46 = 10/23$ for black. I could stack the deck a little more by *replacing* the black cards by red ones; then the chances would be $32/52 = 8/13$ for red and

20/52 = 5/13 for black. Now the odds on the "even" bet are not even at all but are $(8/13)/(5/13)$ = 8/5, or 8 to 5 in my favor. If the black bettor knows the true chances of selection, no problem arises because then the bet can be adjusted to reflect the true odds. The problem arises when the true probabilities are not known.

Stacking the deck occurs in real sampling studies, although (perhaps) not as deliberately as in a gambling environment. To illustrate, what would happen if it was *not known* that the smallest taxpayer was twice as likely to enter the sample as the rest of the taxpayers (Section 10.2)? Clearly, if we *thought* that all taxpayers had an equal chance of entering the sample, then the equally weighted mean $\bar{y}$ would be used. At this point, there would be no reason to use anything else. But in this event, as we saw earlier, $\bar{y}$ is biased; its expectation is 88.0, which underestimates the true mean of 91.50 by 3.5. The deck has been unwittingly stacked in the direction of the smaller taxpayers. Biases of this kind are called selection biases; they occur when the *actual* chances of selection are different from those specified by the design.

A selection bias may occur if the *real* probabilities of selection are different from the *intended* probabilities of selection. If the real probabilities are known, no problem need occur because the known actual probabilities can then be used to form an unbiased estimator. The problem of selection bias occurs if the real probabilities are not known and the incorrect design probabilities are used in the estimator. Thus in the taxpayer illustration, a selection bias occurs if the probabilities specified in the design *and used in the estimator* are equal to 1/8 for all taxpayers, but in reality are 2/9 for taxpayer 1 and 1/9 for all the rest. Then, as we have calculated, a selection bias of $88.0 - 91.5 = -3.5$ occurs.

In practice, selection biases occur with unfortunate frequency. The explanation of the disappearing children in Chapter 1 is selection bias. So perhaps is the unemployment example in Chapter 2. It has been shown in the statistics literature that selection biases can be large in magnitude, subtle, and very difficult to correct. Readers with more advanced interest in selection biases may find additional discussion in the literature of statistics research.

10.3.2 How Bad Can "Good" Data Be?

Measurement errors produce bad data. The sampling procedure itself may be very clever and properly implemented, but measurement errors make the data bad. Earlier we discussed this problem under the title of how bad can *bad* data be?

Unfortunately, data gathered with a selection bias may give every appearance ·of being good. That is, there may be no measurement errors, but the sample may be ruined by selection bias. This is a dishonest sample and leads to the question how bad can *good* data be? Without resorting to mathematics, we state simply that the answer is, "pretty bad."

Since a dishonest sample may contain "good" data with no measurement distortions, detection of a selection bias is very difficult, especially after the sample has been drawn. As a result, it is always important to make the maximum effort to ensure that the sampling operations are carried out *as intended*.

After the sample has been drawn, the best way to detect a selection bias is by comparison with outside data. For example, a sample of the general population should have about 50% males and 50% females. If a selected sample has 75% males and 25% females, the sampling process has almost certainly permitted males to enter the sample with higher probability than females. In the event of such a finding, the appropriate adjustment is to reweight the sample according to the known, correct proportions. To illustrate, if the average weight of males in the sample is 160 lb and of females 120 lb, the adjusted estimate of the population average weight is

$$\frac{1}{2} (160) + \frac{1}{2} (120) = 140 \ lb.$$

This contrasts with the unadjusted estimate of

$$\frac{3}{4} (160) + \frac{1}{4} (120) = 150,$$

which is very much biased towards the heavier males.

Unfortunately, useful outside data are not always available for such comparisons and adjustments, in which case the best warning signal may be peculiarities in the data. The case of the disappearing children is a good example. In that example, the tip off that a selection bias existed was that the number of children seemed to decrease systematically over the three months that each sample area was interviewed.

10.3.3 Coverage

A special case of selection bias occurs when the chance of selection is *zero* for some subgroup of the population. Then the sampled population is *not* the same as the target population. As an illustration, the target population may be the entire United States population, but if the survey is conducted by telephone not every person in the United States can be reached by telephone. Consequently, the sampled population is not the same as the target population. For this reason, the distinction between the *target* population and the *sampled* population is sometimes important.

The statistical inferences that can be made from a sample apply directly only to the sampled population. The extent to which these inferences also carry over to the target population is a matter of judgement and is not necessarily statistically based. Prior to about 1940, the difference between the target population of all United States residents and the telephone population was great. Today, this difference is far less and in many areas of the country is negligible. The result is that telephone surveys are common and that differences between the target and sampled populations are frequently ignored. But the difference between target and sampled populations is not always negligible. In fact, there is an extensive literature on this problem which is referred to as the *coverage* problem.

In summary, selection biases may occur when the actual chances of sample selection are unknowingly different from those specified by the design. Selection biases can be large, subtle, and affect any study. Samplers must always be alert to the possibility that their sample is silently loaded with one particular kind of observation, or possibly systematically missing a unique·identifiable part of the population.

Exercises

10.3.1   One telephone company with a very large number of business customers proposed a sample study of them. Unfortunately, no single list of these customers was available for use as a frame. However the accounting office of the telephone company.did have a list of the "billing numbers" to which separate bills were sent. How many bills any business customer received was determined solely by the way in which that customer wished to be billed. As a result, very similar businesses sometimes received very different numbers of bills, that is, they had a different number of "billing numbers."

Since no list of customers was available, it was proposed that a sample of billing numbers be selected with equal probability and that the business customers associated with those numbers be interviewed. Furthermore, it was suggested that estimates of the total billed amount for all customers be obtained by multiplying the sample totals by N/n where n is the number of customers that appear in the sample and N is the total number of the company's business customers. Discuss this sampling plan. Are the estimates biased or unbiased? Why? How would you improve the scheme?

10.3.2   In household surveys the probability of finding someone at home on a random visit is about 0.7. A sample of size 1000 is to be interviewed.

(a) What is the expected number of household interviews if each household is visited at most three times. Assume that calls are made at random times.

(b) If we want to get at least 99% of the sample households interviewed, how many calls should we expect to make?

(c) In practice would you really expect to interview 99% of the sample in four calls? List the various reasons that you think 99% is unrealistic. What assumption was used in parts (a) and (b) that is contradicted by these reasons?

10.3.3   The following data are taken from a study described by A. L. Finkner in a report "Methods of Sampling and Estimating Commercial Peach Production in North Carolina" published by the North Carolina Agricultural Experiment Station in 1950.

FINKNER MAIL SURVEY

	Number of Growers	Percent of Population	Average No. of Trees per Grower
Response to 1st mailing	300	10	456
Response to 2nd mailing	543	17	382
Response to 3rd mailing	434	14	340
Nonrespondents after 3rd mailing	1839	59	290
Total Population	3116	100	329

(a) If a mail survey of this population was stopped after one mailing, how big a bias would be expected in the estimated number of fruit trees per grower? Calculate this same bias after two mailings, and also after three mailings.

(b) If you were in charge of this survey would you ignore this possibility of bias? If not, what would you do about it?

(c) What other types of surveys can you think of that might also have this difficult property that the ease of response is related to the measurement of interest?

10.3.4   Quota sampling specifies that sampling continues until predetermined sample sizes are obtained in subclasses of the population. For example, sampling might continue until 80 whites and 20 blacks are sampled. A claimed "advantage" for quota sampling is that call-backs are not necessary while they are for other types of sampling. Discuss this claim.

10.3.5   To study various characteristics of corporate stockholders, a proposal was made to randomly select some of the stockholders of publicly owned corporations. Discuss this sampling scheme.

10.4 Systematic Sampling

As we have seen, different ways of gathering up the sample lead to different sampling distributions. So far we are familiar with sampling with equal probability both with and without replacement, and also unequal probability with replacement. The missing case, unequal probability without replacement, is much too complicated for this book. However there is one more sampling method, *systematic* sampling, which we must discuss here even though it too can be complicated.

Familiarity with systematic sampling is important because it is operationally very simple and is therefore very commonly used. The procedure is best illustrated by example. If records are stored on a magnetic tape which is $N = nk$ records long, then a *systematic* sample of size n is selected by picking a random starting point between the first and the kth record, and then selecting every kth record after that. So if the first randomly selected record is the fourth, the sample would consist of the fourth, $(4+k)th, (4+2k)th, \ldots, (4+n-1\ k)th$ records. The total sample size is n.

In practice, there are many places in which systematic sampling is very appealing. Records that are stored in an ordered form on magnetic tape or in card files are easily sampled this way. Systematic sampling is also useful when the population is naturally ordered in time, such as automobiles arriving at a toll booth. The instruction to "take every tenth one" is easy to give and easy to follow.

To illustrate systematic sampling, suppose that a sample of two taxpayers from eight is drawn by selecting a random number between 1 and 4, and then systematically including the fifth, sixth, seventh, or eighth taxpayer. The possible samples are:

$$(y_1, y_5), \quad \bar{y} = 75$$
$$(y_2, y_6), \quad \bar{y} = 87$$
$$(y_3, y_7), \quad \bar{y} = 92$$
$$(y_4, y_8), \quad \bar{y} = 112. \tag{10.17}$$

These four possible means constitute the sampling distribution of $\bar{y}$ obtained by systematic sampling.

In *this* case, the sample mean is unbiased because

$$E(\bar{y}) = \frac{75+87+92+112}{4} = 91.5.$$

But unfortunately, systematic sampling does not always lead directly to unbiased estimators. The difficulty arises when the population size is not a product of the sample size and the systematic interval, $N \neq nk$. As an exercise, consider the problem of drawing a systematic sample of two from the entire population of nine taxpayers. The reader will find that the preceding simple procedure does not have an equally weighted sample mean which is unbiased.

However, a slightly more complicated version of systematic sampling does have an unbiased sample mean. To illustrate, consider again drawing a systematic sample of two taxpayers from nine. To do this associate *two* integers with each of the taxpayers as shown below.

T_1	T_2	T_3	T_4	T_5	T_6	T_7	T_8	T_9
1	3	5	7	9	11	13	15	17
2	4	6	8	10	12	14	16	18

Next select a random number, r, between 1 and 9. The two taxpayers selected into the sample are those associated with the numbers r and $r + 9$. Then it is easily verified that all the possible samples are shown here:

Sample Number

1	2	3	4	5	6	7	8	9

	1	2	3	4	5	6	7	8	9
Taxpayers	T_1	T_1	T_2	T_2	T_3	T_3	T_4	T_4	T_5
Sampled	T_5	T_6	T_6	T_7	T_7	T_8	T_8	T_9	T_9

Since each taxpayer appears in exactly two of the possible samples, the equally weighted mean is unbiased. This same scheme can be used in an analogous manner for different values of n and N.

A difficulty with this more complicated version of systematic sampling is that it is not as practically useful as the simple systematic sampling. The instructions that need to be given are much more complicated, especially if the systematic sample is drawn through time. Fortunately, if N is very large, as it often is in practice, the slight bias in $\bar{y}$ from simple systematic sampling is small and not often significant.

The variance of the systematic sample of 2 from 8 is easily calculated directly in the usual way,

$$\frac{1}{4}\{(75-91.5)^2 + (87-91.5)^2 + (92-91.5)^2 + (112-91.5)^2\}$$

$$= 178.25. \tag{10.18}$$

Systematic sampling has other complications in addition to possible bias. In the systematic sample of taxpayers, the taxpayers are listed in the same order as in Chapter 5. If the taxpayers are put in a different order, different samples and a different sampling distribution result. Unfortunately, if this order is such that the measurements are related to the period of systematic selection, then difficulties arise. These difficulties are not discussed in this book, but a glimpse of the potential is easily understood.

To illustrate with an extreme example, suppose that corporate sales are seasonal, but do *not* change from one year to the next. This means that March sales are the same each year but are not necessarily the same as the other months of the year. Now, if the systematic sampling specifies an observation at a randomly selected month followed by additional observations 12 months later, 24 months later, and so on, then clearly, no matter which month is randomly selected the systematic observations are always the same. This implies that the sample variance is zero and is very misleading because corporate sales do have seasonal changes which would be completely undetected by the systematic sampling procedure. This peculiar result occurs because the selection interval and the data have the same cycle, in this case 12 months. Such seasonal variations are an obvious possibility, but unfortunately systematic behavior in sampled populations is not always so transparent.

There is an important case in which analysis of a systematic sample is simple. If the order of the population is *random* with respect to the measurement, then a systematic sample can be analyzed just as if it *were* a completely random sample! This means that all the easy formulas for simple random sampling can be directly applied to the systematic sample. The assumption of a random order is the key to this kind of analysis; so if a random order can be assumed, or arranged, it is possible to have both the simplicity of operation *and* the simplicity of analysis. Not a bad situation at all.

Exercises

10.4.1   Verify that the sampling scheme (Section 10.4) for systematically selecting two taxpayers from nine by associating two integers with each taxpayer is unbiased for estimation of the population mean.

10.4.2   If a population has $N=11$ units, show how to use the technique of multiple integers described in Section 10.4 to draw a systematic sample of $n=3$. Show that the equally weighted sample mean is unbiased.

10.4.3   Data on computer tapes must be read in the order in which they are stored on the tapes. So systematic sampling is frequently used to sample data stored in this way. Imagine that the nine taxpayers are listed in order on a tape which must be read consecutively and only once. Describe how you would use the multiple integer scheme of Section 10.4 to draw a systematic sample of two taxpayers from nine.

10.4.4   Information was needed by a telephone company on the daily volume of telephone calls in a local area. A proposal was made to obtain a systematic sample of days throughout the year. To do this a random number would be selected between 1 and 14. So if the number picked was 8, then the eighth day of the year would be sampled. Following that the $8 + 14 =$ twenty-second day would be observed and every fourteenth day after that throughout the year.

(a)   Will this sample yield an unbiased estimate of the total yearly volume of telephone calls? Discuss your answer completely.

(b)   Discuss the sampling variance of this plan.

IMPORTANT NEW IDEAS

simple random sampling with replacement   selection bias
random sampling with unequal probability   systematic sampling
sampling with probability proportional to size   weighting observations

CHAPTER 11

THE CLEVER USE OF GROUPS

11.1 Objectives

In the last chapter, we saw that sampling variance can be reduced by unequal probability sampling. Variance can also be reduced by carefully using sampling methods appropriate to the natural grouping of the population. As we shall see, these procedures can be viewed as special cases of unequal probability sampling. Therefore, just as improper use of sampling with unequal probabilities can increase variance, so can improper use of the grouping methods.

We have previously stated that three important design factors which influence variance are: (i) the method of sample selection, (ii) the estimator, and (iii) the sample size. Discussion of methods of sample selection (and the appropriate estimators) was introduced in Chapter 10 and is continued in this one. Estimators that utilize measurements auxiliary to those of main interest are discussed in Chapter 12. Finally, the effect of sample size is discussed in Chapter 13.

11.2 Special Treatment for Special Cases

The special treatment of special cases is the simplest case of the clever use of groups. Actually, this procedure is not new to us because this is exactly what was done in Chapter 5 with the special treatment of the large taxpayer. He constitutes a group of size 1 and the remaining 8 taxpayers constitute a second group. The sampling procedure in the two groups is different. The single member of the first group is always included in the sample, while the members of the second group are included only if randomly selected.

Recall that the original taxpayer population is of size 9. This population is reproduced in Table 11.1 along with the mean and variance of both the complete $(N = 9)$ population and the reduced $(N = 8)$ population formed by excluding the large taxpayer. Both the mean and variance of the reduced population are smaller than those of the complete population. The mean is about 8/9 as large while the variance is less than 1/3 of the variance of the complete population. As we shall see, it is the big difference in the variances which makes it worthwhile giving the special treatment to the big taxpayer.

TABLE 11.1

A TARGET POPULATION OF TAXPAYERS

Taxpayer	Actual Income (y) (Thousands of dollars)	Reported Income (x) (Thousands of dollars)
T_1	60	50
T_2	72	56
T_3	68	66
T_4	94	76
T_5	90	90
T_6	102	100
T_7	116	112
T_8	130	110
T_9	200	175

Complete Population	Reduced Population
$N = 9$	$N = 8$
$\overline{Y}_9 = 103.56$	$\overline{Y}_8 = 91.50$
$\sigma_9^2 = 1621.14$	$\sigma_8^2 = 515.75$

(When ambiguity is unlikely, the subscript notation for the complete and reduced populations is omitted.)

In Chapter 5, the large taxpayer was given special treatment by his deliberate selection as one of the three taxpayers for audit. Since then, we have focused on procedures for selecting the other two taxpayers from the remaining eight. The fact that we originally wanted to estimate the mean of the complete population and not simply the mean of the reduced population has been largely ignored. Now we are going to turn back to the estimation of $\overline{Y}_9$, based on the special sampling scheme which includes the big taxpayer in *every* sample of three taxpayers.

First, notice that the population mean $\overline{Y}_9$ can be written as,

$$\overline{Y}_9 = \sum_{i=1}^{9} y_i/9 \tag{11.1}$$

$$= \frac{8}{9}\,\overline{Y}_8 + \frac{1}{9}\,y_9$$

In the numerical terms of the taxpayers, Expression 11.1 is,

$$\overline{Y}_9 = (60+72+\cdots+200)/9 =$$

$$= \frac{8}{9}\,(91.5) + \frac{1}{9}\,(200) = 103.5. \tag{11.2}$$

In Expressions 11.1 and 11.2, the weights 8/9 and 1/9 are the *known* relative group sizes. Furthermore, the value $y_9 = 200$ is observed for every sample. Consequently, the only unknown in Expression (11.1) for $\overline{Y}_9$ is the mean of the reduced population, $\overline{Y}_8$, which is known only with a complete census of the eight taxpayers. But we know already from the earlier chapters that this reduced mean can be estimated unbiasedly by a sample of two of the remaining eight taxpayers. This suggests using,

$$\overline{y}_s = \frac{8}{9}\,\overline{y}_2 + \frac{1}{9}\,200 \tag{11.3}$$

as an estimator of $\overline{Y}_9$, where $\overline{y}_2$ is the sample mean based on the two (of eight) randomly selected taxpayers. The possible values of $\overline{y}_2$ are of course exactly those that appeared in Table 5.2. They are repeated in Table 11.2. To illustrate, if taxpayers 1 and 6 are randomly selected along with T_9,

$$\overline{y}_s = \frac{8}{9}\,(81.00) + \frac{1}{9}\,(200) = 94.22. \tag{11.4}$$

All the possible values of $\overline{y}_s$ appear in Table 11.2.

This estimator, $\overline{y}_s$, reasonably uses the sample of two taxpayers to represent the eight from whom they are selected and the one large taxpayer to represent himself. Furthermore, $\overline{y}_s$ is unbiased. Since all the possible values of $\overline{y}_s$ appear in Table 11.2, the unbiasedness can be numerically confirmed. It can also be proven mathematically.

As usual, the variance of $\overline{y}_s$ can be computed directly from the sampling distribution. Using Table 11.2,

TABLE 11.2

SAMPLING DISTRIBUTION OF $\bar{y}_s$

(Special Treatment of the Large Taxpayer)

Sample Number	Sampled Taxpayers	$\bar{y}_2$	$\bar{y}_s$	s^2	$var(\bar{y}_s)$
1	1,2	66.0	80.89	72.0	21.33
2	1,3	64.0	79.11	32.0	9.48
3	1,4	77.0	90.67	578.0	171.26
4	1,5	75.0	88.89	450.0	133.33
5	1,6	81.0	94.22	882.0	261.33
6	1,7	88.0	100.44	1568.0	464.59
7	1,8	95.0	106.67	2450.0	725.93
8	2,3	70.0	84.44	8.0	2.37
9	2,4	83.0	96.00	242.0	71.70
10	2,5	81.0	94.22	162.0	48.00
11	2,6	87.0	99.56	450.0	133.33
12	2,7	94.0	105.78	968.0	286.81
13	2,8	101.0	112.00	1682.0	498.37
14	3,4	81.0	94.22	338.0	100.15
15	3,5	79.0	92.44	242.0	71.70
16	3,6	85.0	97.78	578.0	171.26
17	3,7	92.0	104.00	1152.0	341.33
18	3,8	99.0	110.22	1922.0	569.48
19	4,5	92.0	104.00	8.0	2.37
20	4,6	98.0	109.33	32.0	9.48
21	4,7	105.0	115.56	242.0	71.70
22	4,8	112.0	121.78	648.0	192.00
23	5,6	96.0	107.56	72.0	21.33
24	6,7	103.0	113.78	338.0	100.15
25	7,8	110.0	120.00	800.0	237.04
26	6,7	109.0	119.11	98.0	29.04
27	6,8	116.0	125.33	392.0	116.15
28	7,8	123.0	131.56	98.0	29.04
EXPECTATION		91.50	103.50	589.43	174.65

$Var(\bar{y}_2) = 221.04$

$Var(\bar{y}_s = E(\bar{y}_s - \bar{Y}_9)^2$

$\quad = \dfrac{1}{28}[(80.89 - 103.56)^2 + \cdots + (131.56 - 103.56)^2]$

$\quad = 174.65 = (\dfrac{8}{9})^2(221.04)$

$Var(\bar{y}_3) = (1 - \dfrac{3}{9})\dfrac{1823.78}{3} = 405.32$

$$Var(\bar{y}_s) = E(\bar{y}_s - \bar{Y}_9)^2 = \frac{1}{28} [(80.89 - 103.56)^2$$

$$+ \cdots + (131.56 - 103.56)]$$

$$= 174.65$$

As previously, this variance can also be calculated by a mathematically derived formula, specifically,

$$Var(\bar{y}_s) = (\frac{8}{9})^2 Var(\bar{y}_2)$$

$$= (\frac{8}{9})^2 (221.04) \tag{11.5}$$

$$= 174.65.$$

From Expression 11.5, it can be seen that the variance of $\bar{y}_s$ is a multiple of the variance of $\bar{y}_2$ and that the large taxpayer contributes *nothing* to the variance of $\bar{y}_s$! This pleasant result could have been anticipated because the large taxpayer appears in *every* sample of three and as a result does not contribute to the variance of $\bar{y}_s$. But it does, of course, contribute to the estimate itself. Furthermore, since the two randomly selected taxpayers represent only 8/9 of the target population, the standard deviation of $\bar{y}_2$ is multiplied by 8/9 to get the standard deviation of $\bar{y}_s$. Consequently, the standard deviation of $\bar{y}_s$ is always less than the standard deviation of $\bar{y}_2$.

To estimate $Var(\bar{y}_s)$, the estimate $var(\bar{y}_2)$ is substituted for $Var(\bar{y}_2)$ in Expression 11.5. So for example, if taxpayers 1 and 6 are selected, the variance of $\bar{y}_s$ is estimated by

$$var(\bar{y}_s) = (\frac{8}{9})^2 (1 - \frac{2}{8}) \frac{882.0}{2} = 261.33 . \tag{11.6}$$

The estimate of $var(\bar{y}_s)$ is given in Table 11.2 for every sample and it can be readily confirmed numerically that this variance estimation procedure is unbiased. This unbiasedness can also be seen intuitively from the fact that $var(\bar{y}_2)$ is an unbiased estimator of $Var(\bar{y}_2)$ and the multiplier $(8/9)^2$ is a constant which multiplies every possible value of $var(\bar{y}_2)$.

Has the special treatment of the large taxpayer really achieved a reduction in variance? To answer this question fairly, the variance of $\bar{y}_s$ really ought to be compared with the variance of the sample mean that would result if three taxpayers were selected from the nine with equal probability and without replacement, with no special rules for the inclusion of the large taxpayer. Let us call this sample mean $\bar{y}_3$. So in this notation, we want to compare $Var(\bar{y}_s)$ with $Var(\bar{y}_3)$. Since we already know that $Var(\bar{y}_s) = 174.65$ we must now find $Var(\bar{y}_3)$.

In a completely random selection of three taxpayers from nine, there are $\binom{9}{3} = 84$ different samples that could result. These 84 samples appear in Table 11.4. As we might have predicted, $\bar{y}_3$ is unbiased for $\bar{Y} = 103.56$ but the sampling variance of $\bar{y}_3$ is 405.32, which is more than twice the variance (174.65) achieved by the special treatment of the big taxpayer. Clearly, the special treatment of the large taxpayer has reduced the sampling variance substantially.

In this example, the group receiving special treatment consists of only the single large taxpayer, and so the example is somewhat superficial. Nevertheless, in practice where the groups tend to be larger, the technique is equally useful. Usually grouping units which are different from the rest will reduce variance. It is particularly effective if the group isolated for special treatment consists of the "largest" or "smallest" of the elements in the target population.

Finally, as we shall see, the special treatment of population units is really just a special case of a very popular sampling technique called *stratified sampling*.

Exercises

11.2.1   Randomly select three of the numbered samples in Table 11.2 and verify all the calculations given there for those samples.

11.2.2   It is argued that the "smallest" taxpayer T_1 (Table 3.2) should receive special treatment rather than the largest. To evaluate this claim do the following:
(a) Calculate $\bar{Y}$ and S^2 of the reduced population formed by including T_2, T_3, T_4, T_5, T_6, T_7, T_8, T_9.
(b) If T_1 is included in every sample, along with a random sample of two from the reduced population, formulate an unbiased estimator of the mean of the *complete* population.
(c) Calculate the variance of the estimator in (a). (If you are wise you won't do this by calculating all possible values of the estimator.)
(d) Compare the variance in (c) with the variance of the estimator in the sampling scheme (Section 11.2) in which T_9 gets special treatment. Discuss the relative magnitude of the 2 variances.
(e) Compare the variance in (c) with the variance of a simple random sample of 3 from the complete population of all 9

taxpayers. Discuss the comparison.

11.2.3   Using the data given in the text, show that the estimator $var(\bar{y}_s)$ is an unbiased estimator of the sampling variance of $\bar{y}_s$, $Var(\bar{y}_s)$.

11.2.4   The U.S. Department of Commerce reported that in 1968, 1,497,400 companies finished the year with positive assets. The distribution of these assets is given below.

Assets	No. of Companies
Under $100,000	845,900
$100,000 - $999,999	552,900
$1,000,000 - $9,999,999	81,800
$10,000,000 - $24,999,999	9,100
$25,000,000 - $49,999,999	3,500
$50,000,000 - $99,999,999	1,800
$100,000,000 and over	2,300

(a) Calculate the mean value of the assets of these 1,497,400 companies. To do this assume that each company in a given interval has assets equal to the mid-point of the interval. For example, assume that the 845,900 companies with assets under $100,000 each has assets of $50,000. The other mid-points can be taken to be "$550,000, $5,500,000, $17,500,000, $37,500,000, $75,000,000 and $200,000,000.

(b) Using the mid-points given in (a) calculate the variance in assets (σ^2) of this population of companies.

(c) If a sample of 10,000 companies is drawn by simple random sampling from this population of 1,497,400 companies, calculate the sampling variance of the mean value of the assets of the 10,000 companies.

11.2.5   Referring to Exercise 11.2.4 suppose that the 1,800 + 2,300 = 4,100 companies in the two largest dollar intervals are censused and the remaining 5,900 of the 10,000 sample companies are randomly selected from the other five size intervals. In this sampling plan special treatment is being given to the companies with the largest assets.

(a) Calculate the mean value of the assets separately for the companies which are to be sampled and for the companies which are to be censused.

(b) Calculate the variance of the subpopulation of companies in the five smallest asset-intervals.

(c) Assuming this "special treatment" sampling of companies, formulate an unbiased estimator of the mean assets per company for the population of 1,497,000 companies.

(d) Calculate the variance of the estimator in (c) and compare it with the completely random scheme of the previous exercise.

(e) Would it be even more efficient to include the three largest classes of companies in the census group? How about having only the one largest asset-class in the census group? Discuss your answers and support them with numerical calculations.

11.2.6 Various exercises in Sections 3.2, 5.2 and 5.3 concern estimation of the proportion of Democrats in the population of nine taxpayers. This problem is *not* aided by the deliberate inclusion of the large taxpayer in all samples. Why not? Discuss your answer and back it with numerical calculations.

11.2.7 In a marketing study a utility wanted to estimate the average monthly bill of its residential customers. In the particular geographical area under study there were 9050 customers. It was also known that 2210 of the customers received the minimum monthly bill of $8.65.

Two different sample designs were proposed. The first was a random sample of 100 customers from the whole population and the other was a sample of 100 from the 6840 customers who received bills higher than $8.65. Formulate an unbiased estimator of the average monthly bill of all 9050 customers for each of these two sampling plans. Find the sampling variance of the unbiased estimator in each sampling scheme given that the variance of the whole population is 1369 and the variance of the 6840 is 429 (dollars squared).

11.3 A Simple Illustration of Gains and Losses by Grouping

Sampling variance can be reduced by appropriate coordination of the sampling procedure and the grouping of the population units. To illustrate, consider a population that is even simpler than that of the taxpayers; it consists of nine children whose individual ages are 3, 3, 3, 6, 6, 6, 9, 9, 9. Their average age is $\bar{Y} = 6$ years. Also, $\sigma^2 = 6$, and

$S^2 = 27/4$. The problem is to estimate the average age of the nine children by drawing a sample of three of them.

First, if the sample of three is drawn with equal probability without replacement, we know from Chapter 5 that

$$Var\,(\bar{y}) = (1 - \frac{3}{9})\frac{1}{3}\,\frac{27}{4} = 3/2 \qquad (11.7)$$

Confidence limits for the sample mean are based on a sampling distribution with this variance. The question now arises, "Is it possible to sample in such a way that the variance is *less* than 3/2?" It turns out that it is indeed possible.

Consider a different sampling scheme, one which takes advantage of the fact that the children fall into three grades; those in nursery school all of whom are aged 3, those in grade one who are age 6, and those in grade four who are 9. So $G_1 = \{3,3,3\}$, $G_2 = \{6,6,6\}$ and $G_3 = \{9,9,9\}$.

The new sampling scheme specifies that one child be selected with equal probability from each group. So the sample size is still 3, as it was in simple random sampling, but the variance of the new scheme is much different. In the new scheme, *every* possible sample mean has the value, $\bar{y} = (3+6+9)/3 = 6$, which is exactly equal to the true value of $\bar{Y}$. No other result is possible! This means that the estimator not only has zero bias but also has *zero* variance; this is an infinite improvement over simple random sampling!

In the new sampling scheme, use has been made of the fact there is *no* age variance among students in the same grade. All of the variance in age is *among* the grades G_1, G_2, and G_3, and the sampling scheme took advantage of this by drawing a part of the sample from *each* group. Such a scheme is called *stratified sampling* and the groups are referred to as *strata*.

But more might be learned from this example. Suppose it is also known that the nine children are from three families such that $F_1 = \{3,6,9\}$, $F_2 = \{3,6,9\}$, and $F_3 = \{3,6,9\}$. A third possible sampling scheme is to select three children by observing one *complete* family with each family having an equal chance of selection. Then again, no matter which family is selected $\bar{y} = (3+6+9)/3 = 6$. So, this sampling design also results in an estimator which is unbiased and has *zero* variance!

The reason for the zero variance in this case is that there is no variation from family to family, each family has three children with the ages 3, 6, and 9. All of the variability in the population is *within* the families, and the sampling scheme eliminates this by taking *all* of the children within the sampled family. This method of sample selection is called *cluster* sampling for the simple reason that the population units are selected together in clusters, in this case families.

Now we have two sampling schemes, both with zero variance and both clearly better than simple random sampling. However, these sampling schemes are in a sense opposites because one scheme samples all groups and the other samples only one group. This raises the question "What happens if we get mixed up and use cluster sampling with the homogeneous strata and stratified sampling with the heterogeneous clusters?"

First, let's see what happens if *all* of the children are selected from one of the strata G_1, G_2, G_3, rather than one child from each stratum. In this event, the possible sample means are $\bar{y} = 3$ if G_1 is selected, $\bar{y} = 6$ if G_2, and $\bar{y} = 9$ if G_3. Then by direct calculation the variance is,

$$\frac{1}{3} [(3-6)^2 + (6-6)^2 + (9-6)^2] = 6,$$

which is *much* larger than even the variance (3/2) of simple random sampling!

This is a very important lesson. By using a sampling scheme which is not appropriate to the grouped population, we have obtained a variance which not only is larger than the minimum possible (zero in this example), but is much larger (6 compared to 3/2) than simple random sampling.

Second, the same unfortunate result occurs if groups F_1, F_2, and F_3 are sampled by selecting one child from each family rather than selecting the entire family. It can be calculated that the sampling variance is equal to 2, which is not as large as the variance of the previous unfortunate case, but is still larger than simple random sampling.

These last examples show clearly that it is quite easy to construct *bad* sampling schemes. In the remaining sections of this chapter we shall explore sampling techniques based on groups more fully.

Exercises

11.3.1   In the age example of Section 11.3 verify that the variance of the estimator is equal to 2 when the family groups $\underline{F}_1$, $\underline{F}_2$ and $\underline{F}_3$ are sampled by selecting one child randomly from each family (stratified sampling).

11.3.2   Two groups are to be selected with equal probability without replacement from the following four groups. Once a group has been selected into the sample all units within the group are included in the sample.

Groups	Measurements
G_1 :	11, 12, 15
G_2 :	14, 18, 21
G_3 :	23, 29, 31
G_4 :	29, 30, 36

(a) How many different samples can arise with this sampling scheme?
(b) If $\bar{y}$ is the equally weighted mean of the 6 sample observations, write down the sampling distribution of $\bar{y}$.
(c) Show that $\bar{y}$ is unbiased for $\bar{Y}$.
(d) Calculate $Var(\bar{y})$.

11.3.3   Referring to Exercise 11.3.2 if the groups are rearranged as shown in the table below, redo parts (a), (b), (c) and (d) for the new groups. Explain the difference in $Var(\bar{y})$ for the two different arrangements.

G_1 :	11, 18, 31
G_2 :	12, 21, 29
G_3 :	15, 23, 30
G_4 :	14, 29, 36

11.4 Multistage Sampling

11.4.1 Population Grouping

In the last section both stratified and cluster sampling were introduced by the use of a very simple numerical example. It was shown that sampling variance can be reduced by stratified sampling when the population is organized into homogeneous groups and by cluster sampling when the groups are heterogeneous. A homogeneous group is one in which the observations associated with the population elements are very similar to each other. In real populations, the hierarchal groups are not often readily manipulated. As a result, the art of sampling is to pick a sampling scheme to fit the specific characteristics of the target population. To do this, we need to be as familiar as possible with the characteristics of the various selection procedures. In this section multistage sampling is discussed. Cluster sampling which was introduced in the last section is a special case of it.

Multistage sampling is useful when the population units are arranged in hierarchal groups. For example, adults in a United States city may be grouped into city blocks, then into households within city blocks, and then into adults within households. *Three stage sampling* of this population would consist of first randomly selecting a subset of blocks, then within those selected blocks, selecting a subset of households, and finally selecting a sample of adults within the households. In such a scheme, the blocks are referred to as primary sampling units, the households as secondary units, and the adults as tertiary units. It is important to keep the distinction of sampling units in mind because, as we shall see, each of these levels contributes a component to the total sampling variance. Efficient sample design depends heavily on knowledge of the magnitude of these different components. The distinction is also important because sampling units and observational units, which are the units on which the measurements are made, can be confused. The result of such confusion is that variance calculations are mistakenly based on the observational units rather than the sampling units. In the preceding example, the observational units are the adults, and they contribute to variance only in as much as they also happen to be the tertiary sampling units.

11.4.2 Two Stage Sampling of the Taxpayers

To illustrate two stage sampling numerically, suppose that the taxpayer population is divided into the two primaries indicated in the following table. Since we learned in Section 11.3 that it is good to have primaries that are similar to each other, some larger *and* smaller tax-payers have been put into each primary.

$N = 2$ primaries			
	$\underline{P}_1$	$\underline{P}_2$	
Number in Primary	$M_1 = 5$	$M_2 = 4$	$M = M_1 + M_2 = 9$
Taxpayer Incomes	$y_1 = 60$ $y_3 = 68$ $y_5 = 90$ $y_7 = 116$ $y_9 = 200$	$y_2 = 72$ $y_4 = 94$ $y_6 = 102$ $y_8 = 130$	
Total Mean Dispersion	$Y_1 = 534$ $\overline{Y}_1 = 106.8$ $S_1^2 = 3,187.2$	$Y_2 = 398$ $\overline{Y}_2 = 99.5$ $S_2^2 = 574.3$	$Y = Y_1 + Y_2 = 932$ $\overline{Y} = Y/M = 103.5$

The two stage sampling scheme consists of selecting one of the two primaries with equal probability, followed by the selection of three secondaries with equal probability without replacement from that sampled primary. Consequently, the sample size is equal to 3, the same as in the earlier designs. This permits comparisons without introducing confusion due to differing sample sizes.

In general, the sample size selected within the *ith* sampled primary is denoted by m_i with $m = \sum m_i$. Notice that if $m_i = M_i$ for all sampled primaries, multistage sampling then becomes cluster sampling, which was introduced in Section 11.3.

The sampling distribution of this two stage scheme is easy to find. If $\underline{P}_1$ is selected there are $\binom{5}{3} = 10$ different samples that can arise, and if $\underline{P}_2$ is selected there are $\binom{4}{3} = 4$ different samples. So, in all there are 14 different samples, any one of which can arise in this two stage sampling scheme. Furthermore one of these samples must occur, no others are possible. The 14 samples are listed in Table 11.3 along with the equally weighted mean, $\bar{y}$. The other estimators in Table 11.3, $\hat{y}_{ms}$ and $\hat{y}_{msp}$ are discussed later.

The 14 samples in Table 11.3 do not occur with the same relative frequencies. $\underline{P}_1$ is selected with probability 1/2, and after it has been selected there are 10 different equally likely samples that can be selected within $\underline{P}_1$. As a result each sample in $\underline{P}_1$ appears with a relative frequency of $(1/2)(1/10) = 1/20$. By the same argument, each

TABLE 11.3

TWO STAGE SAMPLING OF TAXPAYERS

	Sampled Taxpayers	$\bar{y}$	$\hat{y}_{ms}$	$\hat{y}_{msp}$
	$T_1 T_3 T_5$	72.67	80.74	72.67
	$T_1 T_3 T_7$	81.33	90.37	81.33
	$T_1 T_3 T_9$	109.33	121.48	109.33
	$T_1 T_5 T_7$	88.67	98.52	88.67
Primary 1	$T_1 T_5 T_9$	116.67	129.63	116.67
	$T_1 T_7 T_9$	125.33	139.26	125.33
	$T_3 T_5 T_7$	91.33	101.48	91.33
	$T_3 T_5 T_9$	119.33	132.59	119.33
	$T_3 T_7 T_9$	128.00	142.22	128.00
	$T_5 T_7 T_9$	135.33	150.37	135.33
	$T_2 T_4 T_6$	89.33	79.41	89.33
	$T_2 T_4 T_8$	98.67	87.70	98.67
Primary 2	$T_2 T_6 T_8$	101.33	90.07	101.33
	$T_4 T_6 T_8$	108.67	96.59	108.67
Expectation		103.15	103.56	103.56

sample in $\underline{P}_2$ appears with a relative frequency of $(1/2)(1/4) = 1/8$.

With these relative frequencies, we can see that the equally weighted mean, $\bar{y}$, is a biased estimator of $\bar{Y}$. From Table 11.3

$$E(\bar{y}) = \frac{1}{20} [72.67 + --- + 135.33]$$
$$+ \frac{1}{8} [89.33 + --- + 108.67] \qquad (11.8)$$
$$= 103.15$$
$$\neq 103.56 = \bar{Y}.$$

Clearly, the simple mean is biased. Furthermore, from our discussion of unequal probability sampling in Chapter 10, we know that the source of the bias lies in the unequal relative frequencies of the 14 sample means.

Unbiased estimation requires that the observations be weighted inversely to their probabilities of selection. To do such a weighting we must first find these probabilities. From Chapter 5, we know that in simple random sampling of n from N a particular unit of the population has a chance n/N of appearing in the sample. Since the primaries and secondaries are selected separately by simple random sampling a single primary has a chance n/N of appearing, and within that primary each secondary has a selection chance of m_i/M_i. Together this means that the chance of a single secondary appearing in the sample is the product, $(n/N)(m_i/M_i) = (nm_i)/(NM_i)$. To illustrate, in the IRS example taxpayers in $\underline{P}_1$ appear with a probability of 3/10 and those in $\underline{P}_2$ with a probability of 3/8.

Consequently, if sample $T_1 T_3 T_5$ (Table 11.3) is selected from $\underline{P}_1$,

$$\frac{10}{3} (60) + \frac{10}{3} (68) + \frac{10}{3} (90) = 726.7$$

is a realized value of an estimator of Y, and division by $M = 9$ gives,

$$\frac{1}{9} [10 \frac{(60+68+90)}{3}] = 80.74$$

as a realized value of an estimator of the mean, $\bar{Y}$. Similarly, if sample $T_2 T_4 T_6$ is obtained from $\underline{P}_2$

$$\frac{1}{9} [8 \frac{(72+94+102)}{3}] = 79.41$$

is an estimate of $\bar{Y}$. In both of the above calculations the observations are weighted inversely to the chance of selection. Consequently, we would expect such an estimator to be unbiased. We shall see shortly that this is so.

In general, the estimator introduced previously is,

$$\hat{y}_{ms} = \frac{1}{9} (10\bar{y}), \quad \text{if} \quad \underline{P}_1 \text{ is selected,}$$

and

$$= \frac{1}{9} (8\bar{y}), \quad \text{if} \quad \underline{P}_2 \text{ is selected.}$$

Of course, $\bar{y}$ will differ in the two cases. All of the calculated values of $\hat{y}_{ms}$ appear in Table 11.3. It is easily verified that

$$E(\hat{y}_{ms}) = \frac{1}{20} \ [80.74 + - - - + 150.74]$$
$$+ \frac{1}{8} \ [79.41 + - - - + 96.59] \qquad (11.9)$$
$$= 103.56 = \overline{Y},$$

so that $\hat{y}_{ms}$ is an unbiased estimator of $\overline{Y}$, just as we expected.

In general notation, reciprocal weighting of the observations gives,

$$\sum_i \sum_j \{(NM_i)/(nm_i)\}y_{ij} = \frac{N}{n} \sum_i M_i\overline{y}_i \qquad (11.10)$$

as an unbiased estimator of Y. Then division by the population size, M, gives

$$\hat{y}_{ms} = \frac{1}{M} \frac{N}{n} \sum_i M_i\overline{y}_i, \qquad (11.11)$$

as an unbiased estimator of $\overline{Y}$, where the summations in $\hat{y}_{ms}$ are over the *sampled* primaries and secondaries. The reader should check that the estimator $\hat{y}_{ms}$ used in the taxpayer example is indeed a special case of this general formula.

Sometimes, it is helpful to take a slightly different view of the formula for $\hat{y}_{ms}$. The product $M_i\overline{y}_i$ is an unbiased estimate of the total of the *i*th primary, Y_i, and the average of these products

$$\frac{1}{n} \sum_i M_i\overline{y}_i$$

estimates the average primary total,

$$\frac{1}{N} \sum_{i-1}^{N} M_i\overline{Y}_i = \frac{1}{N} \sum_{i=1}^{N} Y_i.$$

Further, since there are N primaries, multiplication by N gives

$$\frac{N}{n} \sum M_i\overline{y}_1,$$

as an unbiased estimate of $Y = \sum_{i=1}^{N} Y_i$. Finally, division by M gives the estimator $\hat{y}_{ms}$ of the mean $\overline{Y}$. This approach will be used to construct an estimator in Section 11.5.

11.4.3 The Variance in Two Stage Sampling

Now that we know $\hat{y}_{ms}$ is unbiased as an estimator of $\overline{Y}$, what about its variance? First, notice that in the simple random sampling of three taxpayers from nine (Table 11.4) there are 84 possible samples. This compares with the 14 samples that are possible in the two stage sampling scheme (Table 11.3). Furthermore, the 14 samples are a subset of the 84. In Table 11.4 the 14 samples are indicated by the number "1". The two stage sampling scheme has effectively removed the possibility of the other 70 samples occurring. For this reason, multistage sampling is sometimes referred to as *restricted* random sampling. The hope is, of course, that the variance of the estimator will be decreased by removing some of the possible samples. Has such a variance reduction actually occurred in our taxpayer example? Unfortunately, it has not. As we shall see, the unequal primary sizes are creating some difficulty.

Direct calculation of the variance of $\hat{y}_{ms}$ from Table 11.3 gives

$$Var(\hat{y}_{ms}) = \frac{1}{20}\,[(80.74-103.56)^2 + \cdots + (150.37-103.56)^2]$$
$$+ \frac{1}{8}\,[(79.41-103.56)^2 + \cdots + (96.56-103.56)^2]$$
$$= 509.63 \qquad\qquad (11.12)$$

Now an unfortunate thing can be seen. The variance of $\hat{y}_{ms}$ is larger than the variance of $\overline{y}_3$ in simple random sampling. Specifically,

$$Var(\overline{y}_3) = (1 - \frac{3}{9})\,\frac{1823.78}{3} = 405.3 \qquad (11.13)$$

(Table 11.4) compares with $Var(\hat{y}_{ms}) = 509.6$ from Expression 11.12. Regrettably, the variance of the two stage scheme is larger than simple random sampling. Apparently, nothing has been gained and indeed something has been lost. What went wrong?

As we shall confirm later, the estimator $\hat{y}_{ms}$ has the very undesirable characteristic of having a nonzero variance even when the sampled population has *zero* variance. This is a nasty, inefficient feature which is very dangerous in applied studies. As a result, a separate section (11.5) has been devoted to this topic.

As in all the other sampling schemes that we have discussed, the variance of $\hat{y}_{ms}$ can also be calculated by use of a mathematically derived formula. In this case the formula is

TABLE 11.4

SAMPLING DISTRIBUTION OF $\bar{y}$.
(3 taxpayers from 9)

Sample* Number	Sampled Taxpayers	$\bar{y}$	Sample Number	Sampled Taxpayers	$\bar{y}$
1	1 2 3	66.67	43[2]	2 5 9	120.67
2	1 2 4	75.33	44	2 6 7	96.67
3	1 2 5	74.00	45[1]	2 6 8	101.33
4[2]	1 2 6	78.00	46	2 6 9	124.67
5[2]	1 2 7	82.67	47	2 7 8	106.00
6[2]	1 2 8	87.33	48	2 7 9	129.33
7[2]	1 2 9	110.67	49	2 8 9	134.00
8	1 3 4	74.00	50	3 4 5	84.00
9[1]	1 3 5	72.67	51[2]	3 4 6	88.00
10[2]	1 3 6	76.67	52[2]	3 4 7	92.67
11[1,2]	1 3 7	81.33	53[2]	3 4 8	97.33
12[2]	1 3 8	86.00	54[2]	3 4 9	120.67
13[1,2]	1 3 9	109.33	55[2]	3 5 6	86.67
14	1 4 5	81.33	56[1,2]	3 5 7	91.33
15[2]	1 4 6	85.33	57[2]	3 5 8	96.00
16[2]	1 4 7	90.00	58[1,2]	3 5 9	119.33
17[2]	1 4 8	94.67	59	3 6 7	95.33
18[2]	1 4 9	118.00	60	3 6 8	100.00
19[2]	1 5 6	84.00	61	3 6 9	123.33
20[1,2]	1 5 7	88.67	62	3 7 8	104.67
21[2]	1 5 8	93.33	63[1]	3 7 9	128.00
22[1,2]	1 5 9	116.67	64	3 8 9	132.67
23	1 6 7	92.67	65[2]	4 5 6	95.33
24	1 6 8	97.33	66[2]	4 5 7	100.00
25	1 6 9	120.67	67[2]	4 5 8	104.67
26	1 7 8	102.00	68[2]	4 5 9	128.00
27[1]	1 7 9	125.33	69	4 6 7	104.00
28	1 8 9	130.00	70[1]	4 6 8	108.67
29	2 3 4	78.00	71	4 6 9	132.00
30	2 3 5	76.67	72	4 7 8	113.33
31[2]	2 3 6	80.67	73	4 7 9	136.67
32[2]	2 3 7	85.33	74	4 8 9	141.33
33[2]	2 3 8	90.00	75	5 6 7	102.67
34[2]	2 3 9	113.33	76	5 6 8	107.33
35	2 4 5	85.33	77	5 6 9	130.67
36[1,2]	2 4 6	89.33	78	5 7 8	112.00
37[2]	2 4 7	94.00	79[1]	5 7 9	134.33
38[1,2]	2 4 8	98.67	80	5 8 9	140.00
39[2]	2 4 9	122.00	81	6 7 8	116.00
40[2]	2 5 6	88.00	82	6 7 9	139.33
41[2]	2 5 7	92.67	83	6 8 9	144.00
42[2]	2 5 8	97.33	84	7 8 9	148.67
			TOTAL		8698.67
			MEAN		103.56
			VARIANCE		405.32

* "1" denotes two stage sampling and "2" denotes stratified sampling.

$$M^2 Var(\hat{y}_{ms}) = \frac{N^2}{n} (1 - \frac{n}{N}) \frac{1}{N-1} \sum (Y_i - Y^*)^2$$
$$+ \frac{N}{n} \sum \frac{M_i^2}{m_i} (1 - \frac{m_i}{M_i}) S_i^2 , \qquad (11.14)$$

where $Y_i = \sum_j y_{ij}$ and $Y^* = \frac{1}{N} \sum_i Y_i$.

Expression 11.14 clearly shows that the variance of $\hat{y}_{ms}$ is the sum of two variance components, one which reflects differences among the primary totals and another which results from variability among the secondary units within a given primary. In the IRS population, the within primary component of variance is 281.23 and the between primary component is 228.35. This gives a total variance of 509.58.[1] A three stage sampling scheme would have three components of variance.

Exercises

11.4.1 Randomly select three of the samples displayed in Table 11.3 and verify the calculations for $\bar{y}$ and $\hat{y}_{ms}$.

11.4.2 In Section 11.4 in the two stage sampling of taxpayers, it is stated that the between primary variance component is 228.35 and the within primary variance component 281.23. Select the appropriate formulas from Section 11.4 and use them to verify these numerical values.

11.4.3 Verify algebraically that when sampling from the taxpayer population the formula for $\hat{y}_{ms}$ (Expression 11.11) reduces to the formula for $\hat{y}_{ms}$ (subsection 11.4.2).

11.4.4 If the reduced ($N = 8$) taxpayer population is partitioned into the four groups (y_1,y_5), (y_2,y_6), (y_3,y_7), (y_4,y_8), write down the sampling distribution of the equally weighted sample mean of the two observations obtained by selecting one of the four clusters with equal probability. Show numerically that this estimator is unbiased and calculate its sampling variance.

---

1. As a result of rounding error, this total variance differs very slightly from $Var(\hat{y}_{ms})$ = 509.63 which was calculated earlier in Expression 11.12.

11.4.5   Compare the sampling scheme of Exercise 11.4.4 with the systematic sampling of the reduced taxpayer population discussed
in Section 10.5. What is the difficulty associated with the *estimation* of the sampling variance in both of these schemes?

11.4.6   In 1974 Chatham Township, New Jersey, conducted a census
of children under 18. The political area was divided in 100
"blocks" each containing just about 30 households. Among
other things, the number of children per family was observed.
From the census data it was determined that
$\sum (Y_i - Y^*)^2/(N-1) = 36.3$ and $S_i^2 \doteq 4.1$ for all $i = 1,... 100$
blocks, (see Equation 11.14).

It would have been much less work to sample the area
than to census it. If a sample of 25 blocks and 4 households
per block had been selected, each stage with equal probability
without replacement, calculate the variance of the estimator of
the total number of children in Chatham Township. Calculate
the variance of the same estimate if 10 blocks are sampled
with 10 households per block.

If the fixed cost of processing the data is $300 and the
variable cost is $1.50 per household, calculate the total cost of
processing the data for the census and the two sampling
schemes given above.

If it costs $10 to travel to each block and $10 to interview
each family, calculate the interviewing cost of the two sampling plans given above. Also calculate the interviewing costs
of the census.

Calculate the total cost of the census and the two sampling
plans. Discuss your results.

11.4.7   In Table 11.4 verify that the indicated samples are actually the
40 permitted by stratified sampling and the 14 permitted by
two stage sampling.

11.5 Sampling Unequally Sized Groups

In the previous sections, the gains in efficiency made possible by
the clever coordination of the sampling scheme and the grouping of the
target population were discussed. But in Section 11.43, we observed
that the estimator discussed there, $\hat{y}_{ms}$, has a variance larger than the
variance of simple random sampling. As we shall see, this increase in

variance comes directly from variability in the primary sizes. This difficulty and methods for handling it are discussed in this section in some detail.

11.5.1 A Population with Zero Variance

We begin with a very simple example. Suppose that three primary groups contain two, four, and six secondary units, respectively, and that *every* element in the population is 1. Then, in the multistage notation of Section (11.4),

$$M = M_1 + M_2 + M_3 = 2 + 4 + 6 = 12,$$

$$Y = Y_1 + Y_2 + Y_3 = 2 + 4 + 6 = 12,$$

and

$$\overline{Y} = \overline{Y}_1 = \overline{Y}_2 = \overline{Y}_3 = 1. \tag{11.15}$$

Since each measurement in the population is the same, the population has zero variance. As a result, any sampling scheme that does *not* have zero variance must be judged to be very poor; after all, the sampling scheme will have introduced variability when none exists in the population!

Suppose that the preceding zero-variance population is sampled by selecting one of the three primaries with equal probability, 1/3. Keep in mind that this is the same procedure that was used to select primaries in the previous section. In this example how, or how many, secondaries are sampled within the selected primary does not matter because the secondary sample mean is always equal to 1. This follows because all the observations are equal to 1. As a result, it is easy to calculate the values of the estimates.

If $\underline{P}_2$ is selected, an estimate of the primary total can be obtained by multiplying the secondary mean (which is always equal to 1) by the number of elements in $\underline{P}_2$ to get $4 \times 1 = 4$, as an estimate of the primary total. Then, since $\underline{P}_2$ was selected with probability 1/3, reciprocal weighting gives $3 \times 4 = 12$ as an unbiased estimate of the population total. At this point no problems are apparent because the estimate is exactly equal to the true value of 12. This is very nice but before forming any strong positive opinions about the sampling scheme, let's calculate the estimates that would be obtained if $\underline{P}_1$ or $\underline{P}_3$ is sampled.

If $\underline{P}_1$ is selected, any selected secondary sample mean from within $\underline{P}_1$ will be equal to 1, just as in $\underline{P}_2$. Furthermore, using exactly the same reasoning as before and the fact that there are two secondaries in $\underline{P}_1$, $2\times1 = 2$ estimates the total of P_1 and $3\times2 = 6$ estimates the population total. This estimate is *not* equal to the true population total of 12.

Finally, if $\underline{P}_3$ is selected, the estimate of the primary total is 6×1 and the estimate of the population total is $3\times6 = 18$. Table 11.5 gives the sampling distribution of the estimators of the total and mean of this very simple population. The estimator of the mean is obtained by division by the population size, 12.

TABLE 11.5

SAMPLING DISTRIBUTION IN A ZERO
VARIANCE POPULATION WITH UNEQUALLY SIZED PRIMARIES

Primary Selected	Estimate of Total	Estimate of Mean
$\underline{P}_1$	6	1/2
$\underline{P}_2$	12	1
$\underline{P}_3$	18	3/2

Since each of the values 6, 12, 18 in Table 11.5 is equally likely to occur, the expectation of the estimator of the total is equal to $1/3(6+12+18) = 12$. Consequently, the estimator is unbiased. Next, the sampling variance of this estimator is equal to

$$(1/3)[(6-12)^2 + (12-12)^2 + (18-12)^2] = 24.$$

Clearly, this is an unfortunate result - the estimator has a positive variance while at the same time the sampled population has zero variance!

This incredible insertion of inefficiency is a result of the *equal* chance selection of groups of *unequal* sizes. This is perhaps most easily seen by writing the estimator in terms of the primary sizes M_1, M_2 and M_3, as follows. Then, the estimator equals

TABLE 11.6

SAMPLING DISTRIBUTION OF $\bar{y}_{st}$

Sample Number	Sampled Taxpayers	$\bar{y}$	$\bar{y}_{st}$
1	1 2 6	78.00	82.00
2	1 2 7	82.67	88.22
3	1 2 8	87.33	94.44
4	1 2 9	110.67	125.56
5	1 3 6	76.67	80.89
6	1 3 7	81.33	87.11
7	1 3 8	86.00	93.33
8	1 3 9	109.33	124.44
9	1 4 6	85.33	88.11
10	1 4 7	90.00	94.33
11	1 4 8	94.67	100.56
12	1 4 9	118.00	131.67
13	1 5 6	84.00	87.00
14	1 5 7	88.67	93.22
15	1 5 8	93.33	99.44
16	1 5 9	116.67	130.56
17	2 3 6	80.67	84.22
18	2 3 7	85.33	90.44
19	2 3 8	90.00	96.67
20	2 3 9	113.33	127.78
21	2 4 6	89.33	91.44
22	2 4 7	94.00	97.67
23	2 4 8	98.67	103.89
24	2 4 9	122.00	135.00
25	2 5 6	88.00	90.33
26	2 5 7	92.67	96.56
27	2 5 8	97.33	102.78
28	2 5 9	120.67	133.89
29	3 4 6	88.00	90.33
30	3 4 7	92.67	96.56
31	3 4 8	97.33	102.78
32	3 4 9	120.67	133.89
33	3 5 6	86.67	89.22
34	3 5 7	91.33	95.44
35	3 5 8	96.00	101.67
36	3 5 9	119.33	132.78
37	4 5 6	95.33	96.44
38	4 5 7	100.00	102.67
39	4 5 8	104.67	108.89
40	4 5 9	128.00	140.00
TOTAL		3874.67	4142.22
EXPECTATION		96.87	103.56
VARIANCE		186.32	300.43

$$3M_1 \quad \text{if} \quad \underline{P}_1 \text{ is selected,}$$
$$3M_2 \quad \text{if} \quad \underline{P}_2 \text{ is selected,} \qquad (11.16)$$
$$\text{and } 3M_3 \quad \text{if} \quad \underline{P}_3 \text{ is selected.}$$

This makes it very clear that the estimator can vary simply as a result of the differences in $M_1, M_2,$ and M_3. There would be no variation if $M_1 = M_2 = M_3$.

What can be done about this? The obvious suggestion is to arrange groups of equal size. Unfortunately, this cannot always be accomplished in practice. But curiously and pleasantly, the same effect can be achieved by mathematical methods, specifically by selecting the primaries with probabilities which are proportional to their sizes. In this example, the relative sizes of $\underline{P}_1, \underline{P}_2,$ and $\underline{P}_3$ are $2/12 = 1/6$, $4/12 = 1/3$, and $6/12 = 1/2$, respectively. This means that $\underline{P}_1$ is selected with probability $1/6$, $\underline{P}_2$ with probability $1/3$, and $\underline{P}_3$ with $1/2$. Then weighting inversely to these probabilities implies the following estimates of the population total:

$$6{\times}2 = 12 \text{ when } \underline{P}_1 \text{ is sampled,}$$
$$3{\times}4 = 12 \text{ when } \underline{P}_2 \text{ is sampled,}$$
$$\text{and } 2{\times}6 = 12 \text{ when } \underline{P}_3 \text{ is sampled.} \qquad (11.17)$$

This estimator is much better. Not only is it unbiased, but it is always equal to the true value of 12 and so has zero variance. No undesirable variability has been introduced by the unequal primary sizes.

11.5.2 Unequally Sized Groups of Taxpayers

Since sampling the primaries with probabilities proportional to size helped in the example of Section 11.5.1 what is the result of using this technique in the taxpayer population? Since $\underline{P}_1$ has five secondaries and $\underline{P}_2$ has four, sampling with probability proportional to size means that $\underline{P}_1$ is selected with probability $5/9$ and $\underline{P}_2$ with probability $4/9$. As a result, the relative frequency of the 10 samples possible from $\underline{P}_1$ is $(5/9)(1/10) = 1/18$ and the relative frequency of the 4 samples possible from $\underline{P}_2$ is $(4/9)(1/4) = 1/9$. These findings can be used to construct an unbiased estimator and to find its variance.

First, by reciprocal weighting,

$$(\frac{9}{5}) \, 5\bar{y} = 9\bar{y} \quad \text{if} \quad \underline{P}_1 \text{ is sampled}$$

and

$$(\frac{9}{4})\ 4\bar{y} = 9\bar{y} \quad \text{if } \underline{P}_2 \text{ is sampled}$$

is an unbiased estimator of the population total. Division by the population size, $M = 9$, gives

$$\hat{y}_{msp} = \bar{y} \tag{11.18}$$

as an unbiased estimator of the population mean $\bar{Y}$. This estimator refers to multistage sampling with probability proportional to size and to emphasize this the estimator has been given the special notation $\hat{y}_{msp}$.

The complete sampling distribution of $\hat{y}_{msp}$ is given in Table 11.3. Direct calculation of the expectation of $\hat{y}_{msp}$ gives

$$\begin{aligned}
E(\hat{y}_{msp}) &= \frac{1}{18}\ [72.67+\cdots+135.33] \\
&+ \frac{1}{9}\ [89.33+\cdots+108.67] \\
&= 103.56
\end{aligned} \tag{11.19}$$

which verifies that $\hat{y}_{msp}$ is unbiased. This was expected by the way in which the estimator was constructed. In addition,

$$\begin{aligned}
Var(\hat{y}_{msp}) &= \frac{1}{18}\ [(72.67-103.56)^2 +\cdots+ (135.33-103.56)^2] \\
&+ \frac{1}{9}\ [(89.33-103.56)^2 +\cdots+ (108.67-103.56)^2] \\
&= 270.50.
\end{aligned} \tag{11.20}$$

This variance calculation permits us to see the effect of the unequal primary sizes because sampling with probability proportional to size has reduced the variance from 509.6 to 270.50. This can be checked by comparing Expressions 11.12 and 11.20. This comparison is not misleading in practice where the inflation of variance due to varying primary sizes can be substantial. Finally, recall that two stage sampling with equal probability selection of primaries has a variance that is *larger* than simple random sampling. However, by using probability proportional to size, two stage sampling has a variance which is smaller than simple random sampling, 270.50 versus 405.32; see Expression 11.20 and Table 11.2.

It is also possible to remove variability caused by unequal primary sizes by the clever use of a special technique called ratio estimation. However, while ratio estimation is discussed in Chapter 12, its use in multistage sampling has not been included in this book.

Exercises

11.5.1   Verify that the formula for $\hat{y}_{ms}$ Expression 11.11 reduces to the formula for $\hat{y}_{ms}$ developed in subsection 11.5.1 (Expression 11.16).

11.5.2   Randomly select three samples from Table 11.3 and verify the calculations given there for $\hat{y}_{msp}$.

11.6 Stratified Sampling

Stratified sampling was introduced in Section 11.3. A sampling process is called *stratified* if the entire population is divided into mutually exclusive groups and some population units are randomly selected from *every* group. The groups are referred to as *strata*. In Section 11.3, we learned that stratified sampling is most effective when the units within the individual strata are very much alike. In this section, the characteristics of stratified sampling are more fully explored in the context of the now-familiar taxpayer population.

11.6.1 Stratification of the Taxpayers

Suppose that the taxpayers are divided into the two strata given in the following table. Since homogeneous strata are most effective in reducing variance, these two strata are formed by placing the bigger taxpayers in one stratum and the smaller ones in the other. It is important to understand that they are *not* formed in the same way as the primaries of Section 11.4.2. Those primaries were formed by placing some of the big *and* small taxpayers in each primary.

To draw a stratified sample, we select 2 units from $\underline{S}_1$ and 1 from $\underline{S}_2$ with equal probability without replacement. This plan has a sample size of 3, the same as the earlier sampling plans. How many such samples of size 3 are there? First, there are $\binom{5}{2} = 10$ samples that can be obtained from $\underline{S}_1$ and similarly there are $\binom{4}{1} = 4$ from $\underline{S}_2$. Further, since any sample of size 2 from $\underline{S}_1$ can appear with any of the samples of size 1 from $\underline{S}_2$, there are $10 \times 4 = 40$ different samples of size 3 that can arise. The 40 different samples are given in Table 11.6.

We saw earlier that multistage sampling can be viewed as restricted random sampling in the sense that multistage sampling precludes some of the samples that are possible with simple random sampling; see Table 11.4. The hope is that the restricted sampling eliminates some of the more variable samples. Stratified sampling is another example of

$N = 2$ strata			
	$\underline{S}_1$	$\underline{S}_2$	
Number in Stratum	$M_1 = 5$	$M_2 = 4$	$M = M_1+M_2 = 9$
Taxpayer Incomes	$y_1 = 60$ $y_2 = 72$ $y_3 = 68$ $y_4 = 94$ $y_5 = 90$	$y_6 = 102$ $y_7 = 116$ $y_8 = 130$ $y_9 = 200$	
Total Mean Dispersion	$\underline{Y}_1 = 384.0$ $\overline{Y}_1 = 76.8$ $S_1^2 = 213.2$	$\underline{Y}_2 = 548.0$ $\overline{Y}_2 = 137.0$ $S_2^2 = 1894.7$	$Y = Y_1+Y_2 = 932$ $\overline{Y} = Y/M = 103.5$

restricted random sampling. In this case, stratification reduces the number of possible samples from 84 in simple random sampling to 40. For easy comparison the 40 samples possible by stratified sampling are identified in Table 11.4 along with the 14 two stage samples and the 84 of simple random sampling.

Each of the 40 samples in Table 11.6 occurs with the same relative frequency. This is easily seen. From simple random sampling we know that any sample in stratum one occurs with chance 1/10 and that any sample in stratum two occurs with chance 1/4. Further, since any sample from $\underline{S}_1$ can occur with any sample from $\underline{S}_2$, the complete sample occurs with a relative frequency of $(1/10)(1/4) = 1/40$. This result enables us to calculate that the equally weighted mean $\bar{y}$ has expectation

$$E(\bar{y}) = \frac{1}{40} [78.00 + \cdots + 128.00]$$
$$= 96.87 \neq \overline{Y} .$$
(11.21)

So we see that the equally weighted mean is biased in stratified sampling.

The reason for this bias is that stratified sampling does not always give each population unit an equal chance to enter the sample and certainly does not in this example. To see this, observe that, within S_1, the chance that the first taxpayer enters the sample is $m_1/M_1 = 2/5$. This can be verified directly by counting (Table 11.6) that T_1 appears in 16 of the 40 equally likely samples. Also in $\underline{S}_2$, each taxpayer has an $m_2/M_2 = 1/4$ chance of actually occurring, which can also be verified by direct counting. For example, taxpayer 6 appears in 10 of the 40 possible samples. In general, m_i/M_i is the overall probability that a particular secondary within the _ith_ stratum is drawn into the sample. This probability is not necessarily the same for every stratum.

Since the population units do not have an equal chance to enter the sample, unequal weighting is necessary to achieve unbiased estimators. As we know from our earlier work, this unequal weighting is inversely proportional to the chance of selection. So, each observation must be weighted by 5/2 in stratum one, and by 4/1 in stratum two. In general, this weighting is M_i/m_i. So for example, if sample number 1 in Table 11.6 with taxpayers 1, 2, and 6 is selected,

$$\frac{5}{2}\ (60) + \frac{5}{2}\ (72) + \frac{4}{1}\ (102) = 738$$

is a particular value of an unbiased estimator of the total Y. Then division by the population size, 9, gives,

$$\bar{y}_{st} = \frac{5}{9}\ (\frac{60+72}{2}) + \frac{4}{9}\ (102) = 82.0 \tag{11.22}$$

as a particular value of an unbiased estimator of the $\bar{Y}$. This unbiasedness can be confirmed (using Table 11.6) by the calculation,

$$E(\bar{y}_{st}) = \frac{1}{20}\ [82.00 + \cdots + 140.00]$$

$$= 103.56 = \bar{Y}\ . \tag{11.23}$$

In general, the formula for unbiased estimation of $\bar{Y}$ is

$$\bar{y}_{st} = \sum_{i=1}^{N} \frac{M_i}{M}\ \bar{y}_i\ , \tag{11.24}$$

where $\bar{y}_i$ denotes the equally weighted mean within the _ith_ stratum.

The next question is, "What is the variance of $\bar{y}_{st}$?" As usual, this can be calculated directly or by use of a mathematically derived formula. From Table 11.6, direct calculation gives

$$Var(\bar{y}_{st}) = \frac{1}{40}\ [(82.00 - 103.56)^2 + \cdots + (140.00 - 103.56)^2]$$

$$= 300.43. \tag{11.25}$$

This variance is less than the variance of $\bar{y}$ in simple random sampling, Specifically, this variance, 300.43, compares with 405.32, the variance of $\bar{y}_3$ in simple random sampling, see Table 11.2. The decrease in variance occurs because the stratified sampling has eliminated some of the samples that are possible in unrestricted random sampling and which tend to increase variance. In practice, the use of naturally occurring strata seems to result in about a 20% reduction in variance when compared to simple random sampling.

In the previous paragraph we referred to a formula for the sampling variance of $\bar{y}_{st}$. Specifically, this formula is

$$Var(\bar{y}_{st}) = \sum \left(\frac{M_i}{M}\right)^2 \left(1 - \frac{m_i}{M_i}\right) \frac{S_i^2}{m_i}, \qquad (11.26)$$

where

$$S_i^2 = \frac{1}{M_i - 1} \sum_{j=1}^{M_i} (y_{ij} - \bar{Y}_i)^2.$$

The formula shows that the variance of $\bar{y}_{st}$ is the sum of variances *within* each of the strata. This observation confirms the conclusion which was drawn from the age example (11.3) that stratified sampling is most effective when the units within each stratum are similar to each other. This similarity makes individual values of S_i^2 small. The reader can easily confirm that calculations of $Var(\bar{y}_{st})$ by the formula in Expression 11.26 gives the same result as Expression 11.25.

To estimate $Var(\bar{y}_{st})$ use the sample value,

$$s_i^2 = \frac{1}{m_i - 1} \sum_j^{m_i} (y_{ij} - \bar{y}_i)^2,$$

in place of S_i^2. Notice that since the factor $(m_i - 1)$ appears in the denominator of s_i^2, each stratum needs at least two observations to permit calculation of s_i^2. This also implies that the total sample size must be bigger than $2N$. In the taxpayer example, the sampling variance of $\bar{y}_{st}$ cannot be estimated because only one observation was taken from the second stratum. But this difficulty does not harm the understanding of stratified sampling that we have developed in this section.

11.6.2 Sample Allocation

So far we have given no guidance in how many sample units should be allocated to each stratum. This is an important question. If a total of m units are to be observed, how many should be observed in each stratum? It can be shown mathematically that the allocation which minimizes the variance of $\bar{y}_{st}$ for a fixed total sample size is given for the *ith* stratum by the following formula,

$$m_i^{opt} = \frac{M_i S_i}{\sum M_i S_i} m. \qquad (11.27)$$

This formula allocates more units to those strata which are larger (M_i) and/or are more variable (S_i). As a result, we may immediately suspect that the allocation of 2 units to stratum one and 1 to stratum two may not be the best we can do in the taxpayer population. The reason for this suspicion is that stratum one is only 25% larger in size than stratum two but stratum two has a much larger variance, specifically 213.2 to 1894.7. Calculation of the formula for the optimum allocation bears out the suspicion because

$$m_1^{opt} = \frac{5 \times \sqrt{213.2}}{5 \times \sqrt{213.2} + 4\sqrt{1894.6}} (3) = 0.89 \ . \qquad (11.28)$$

Since only integer sample values are possible, the formula suggests that one unit be assigned stratum one. Then by subtraction,

$$m_2^{opt} = 3 - m_1^{opt} = 2.$$

The formula for optimum allocation tells us that the sampling variance of $\bar{y}_{st}$ is less if we use $m_1 = 1$ and $m_2 = 2$ instead of the actual allocation of $m_1 = 2$ and $m_2 = 1$. Fortunately, we do not need to construct another whole sampling distribution to verify this. All that we need do is calculate $Var(\bar{y}_{st})$ by using the formula for it (Expression 11.26), specifically,

$$Var(\bar{y}_{st}) = (\frac{5}{9})^2(1 - \frac{1}{5}) \frac{213.2}{1} + (\frac{4}{9})^2(1 - \frac{2}{4}) \frac{1894.6}{2}$$

$$= 146.21. \qquad (11.29)$$

In summary, if 2 sample units are allocated to stratum one and 1 to stratum two $(m_1 = 2, \ m_2 = 1)$, then $Var(\bar{y}_{st}) = 300.43$. In contrast, optimum allocation of $m_1 = 1$ and $m_2 = 2$ gives $Var(\bar{y}_{st}) = 146.21$. The optimum variance is less than one-half of the variance achieved by stratified sampling with the nonoptimum choice of stratum sample sizes. This example suggests that optimum allocation may be a useful practical procedure.

Regrettably, actual use of the formula for optimum allocation is not always easy. The first problem is that S_i^2 is rarely known in practice and so cannot be substituted in the formula. Furthermore, even an estimate of S_i^2 may not be available in advance of sampling, and this difficulty alone can turn calculation of an optimum allocation into a guessing game. The second problem is that the formula applies only to a *single* measured item while real surveys involve many measurements. Each individual measurement has its own optimum allocation, and not surprisingly, these different allocations are different for the different measurements. Which one to use is not always obvious. There are theoretical procedures for resolving this dilemma but these procedures are mostly artificial and unsatisfactory. A result of these difficulties is that allocation which is proportional to the strata sizes is used far more than optimum allocation. The formula for proportional allocation is

$$m_i^{prop} = \frac{M_i}{M} \ m \qquad (11.30)$$

for all strata. Proportional allocation is discussed in Section 11.8.

11.6.3 The Terminology of Primaries and Strata

Before finishing the section on stratified sampling it is helpful to discuss the terminology of "primaries" and "strata." In both Section 11.4 on multistage sampling and this one, the taxpayer population is divided into two groups. In one case the groups are called primaries and in the other they are called strata. The key to the difference in terminology is that in stratified sampling some observations are made in *every* group; then the groups are called strata. However, if observations are not taken from within every group, but rather only some groups are to be sampled, then the groups are then called primaries. Clearly, the multistage sampling and stratified sampling are very closely related; in fact when $n = N$, the formula for $Var(\bar{y}_{ms})$, Expression 11.14, is *exactly* equal to the formula for $Var(\bar{y}_{st})$ in Expression 11.26.

How is the terminology used if the population is in a three stage hierarchy such as the blocks, households, adults example discussed in Section 11.41? If each of the three levels is *sampled* (as opposed to an exhaustive census), the groups are referred to as primaries, secondaries, and tertiaries. However, if observations are taken from *every* block, then it is customary to refer to the blocks as *strata*, the households as *primaries*, and the adults as secondaries. In the first case the blocks are primaries and in the second case the households are primaries. This usage may appear to be inconsistent, but fortunately it is not as inconsistent as it may first appear to be. The guiding principle is that the terminology "primaries" is reserved for the highest groups in the hierarchy at which proper sampling, as opposed to a census, really occurs.

Exercises

11.6.1   Randomly select three of the samples displayed in Table 11.6 and verify the calculations given there for each of the three samples.

11.6.2   In Section 11.2 the largest taxpayer was given special treatment by his inclusion in every sample. This special treatment scheme is equivalent to stratified sampling in which one stratum contains exactly one taxpayer and the other stratum contains the remaining eight. Viewing the sampling scheme of Section 11.2 as stratified sampling, calculate
(a) the strata totals, Y_1 and Y_2,
(b) the strata means, $\bar{Y}_1$ and $\bar{Y}_2$,
(c) the within stratum variances, S_1^2 and S_2^2.
(d) $Var(\bar{y}_{st})$ and verify that it is equal to 174.65, the same

value that was calculated in Section 11.2.

11.6.3　As in Exercise 11.6.2 view the special treatment of the large taxpayer as a stratified sampling scheme and optimally allocate a sample of size $m = 3$ to the two strata.

11.6.4　The results of the previous exercise suggest allocating no sample units to the stratum with the large taxpayer. If the large taxpayer is not sampled, show that the sample mean has a bias of -12.06 as an estimator of $\bar{Y} = 103.56$. [Some helpful calculations are given in Table 12.1.] Describe this bias verbally.

11.6.5　In subsection 11.6.1, a random sample of two was drawn from the taxpayer stratum $\underline{S}_1$, and a random sample of one from the other stratum, $\underline{S}_2$. Suppose that the two strata, $\underline{S}_1$ and $\underline{S}_2$ are "collapsed" and that the total sample of three is regarded as a random sample of three from the "collapsed" stratum of size nine.
　(a) Is this sample of three equivalent to a random sample of three taxpayers drawn with equal probability and without replacement from the nine in the "collapsed" stratum? [Hint: Compare the sampling distributions.]
　(b) Is the equally weighted mean, $\bar{y}$, of the three observations a unbiased estimator of the mean $\bar{Y} = 103.56$?
　(c) What is the sampling variance of $\bar{y}$?
　(d) Compare the variance calculated in (c) with the sampling variance of $\bar{y}$ when three taxpayers are selected from nine with equal probability without replacement. [The easiest way to calculate this last variance is by the appropriate formula, not by considering all possible samples.]

11.6.6　Referring to Exercise 11.6.5, describe a method for discarding one of the sampled units in such a way that the remaining two units *do* form a random sample of two from the collapsed stratum.

11.6.7　In stratified sampling (Table 11.6),
　(a) Why does the equally weighted mean, $\bar{y}$, have a smaller sampling variance than $\bar{y}_{st}$?
　(b) Calculate and compare the mean square errors of the two estimators in (a).
　(c) Do you think that the larger mean square error of $\bar{y}_{st}$ is too high a price to pay for an unbiasedness?

11.6.8　Use the *formula* for $Var(\bar{y}_{st})$ to verify that the variance of $\bar{y}_s$ is equal to 174.65, as given in Table 11.2.

11.6.9　In the manufacture of electronic circuitry, "wafers" are produced in batches. From a day's production, 10 batches each containing 100 "wafers" are randomly selected and 4, 0, 4, 3, 0, 0, 0, 2, 3, 0 defective wafers are observed in each of the individual batches respectively. Estimate the sampling variance of the observed proportion of defectives, $p = 0.016$. Ignore finite population corrections.

11.6.10 For 1968 the Department of Commerce reported that in the states of Alabama, Georgia, New Jersey and New York labor unions had 158,000, 166,000, 701,000, and 2,453,000 members, respectively.

(a) If a sample of 1200 members is to be allocated to these four states, calculate the proportional allocation.

(b) If the standard deviation of annual earnings, S_i, within the states is \$4000, \$3800, \$2100 and \$1900 respectively, calculate the optimum allocation of the sample of 1200 to the four states.

(c) Calculate the variance of the stratified estimator of average earnings for each of the sample allocations given in (a) and (b). Ignore finite population corrections. Calculate the proportional reduction in variance of the optimum allocation over the proportional allocation.

11.6.11 In reference to Exercise 11.4.6, if a stratified sample is used with one household selected from each block, calculate the variance and total cost of the scheme. Compare these results with those obtained in Exercise 11.4.6.

11.6.12 Answer the questions listed below for the following population. Ignore finite population corrections.

	Stratum			
	1	2	3	4
M_i	200	400	300	100
$\overline{Y}_i$	8	4	4	6
S_i^2	9	4	4	9

(a) Compute the variance of the stratified estimator $\overline{y}_{st}$ assuming proportional allocation of a sample of 100.

(b) Compute the variance of the stratified estimator assuming an equal allocation of 25 to each stratum.

(c) Compute the sampling variance of the sample mean $\overline{y}$ obtained from a completely random sample of size 100.

(d) If strata 1 and 2 and strata 3 and 4 are combined into two new "large" strata and proportional allocation of 100 units is used, compute the variance of $\overline{y}_{st}$. Compare this variance with the variances in parts a, b and c.

11.6.13 A population consists of three groups of sizes 10, 20 and 30 so that the total population size is 60. A sample of size six is drawn from this population in the ways listed below. For each of them, how many different samples of size six are possible? Assume sampling without replacement.
(a) A completely random sample of six from 60.
(b) A sample of size two from each of the three strata.
(c) A stratified sample with proportional allocation.

11.6.14 For sampling scheme (a) of Exercise 11.6.13, what is the probability that a given unit of the population will enter the sample? Answer the same question for schemes (b) and (c) for each of the three strata separately.

11.6.15 In earlier exercises in Sections 3.2, 5.2 and 5.3, we saw that $S^2 = NPQ/N - 1$ where $P(Q = 1 - P)$ is the proportion of the population possessing a certain characteristic (for example registered Democrats). So for two large strata of equal size the formula for optimization allocation becomes

$$m_{opt}^i = \frac{\sqrt{P_i Q_i}}{\sum_i \sqrt{P_i Q_i}} \, m \, , \quad i = 1, 2.$$

(a) Verify the above formula.
(b) If one stratum is thought to have about 1/2 Democrats and the other stratum about 3/4 Democrats which one should be allocated more sampling units?

11.7 Variance Considerations

In two stage sampling, the variance formula has two components, one of which accounts for variability among primaries, and the other for variability among secondaries within primaries. We saw this in Section 11.4. Similarly, three stage sampling has three variance components, one associated with each of the three levels of sampling. In general, a k-level hierarchy would have k variance components.

In most applications, the largest component of variance is associated with the primary sampling units. The variance components associated with secondaries and tertiaries are nearly always smaller in size. In fact, in a three stage scheme, sometimes the secondary and tertiary components *together* do not total more than 10 to 15% of the total variance. A practical result of this is that approximate variances are occasionally calculated by ignoring the lower level variances entirely! However, there is a more important point to this observation.

If the primary variance component is the largest, the sample should be spread out as much as possible over the primary sampling units. In the three stage sampling of blocks, households, and adults, the biggest component of variance will usually be among blocks. Consequently, the sample should cover as many blocks as possible, even at the expense of a reduction in the number of households selected within each block. The best procedure, of course, is to have some observations from *every* block[2] because then the finite population correction for the block sampling is equal to zero and eliminates the block variance completely. In that event, the dominant variance component is the variance among households, which could reasonably be expected to be very much smaller than the variance among blocks. Such a substantial reduction in variance is well worth pursuing. In fact, if there are many strata, some knowledgeable practitioners will actually assign *a single* observation to each stratum because such an extreme design can remove a substantial amount of variance. But before such an extreme allocation is actually used, recall from the section on stratified sampling (11.6) that a single observation per stratum creates a problem of variance estimation. This problem has to be dealt with by difficult estimation techniques which are not covered in this book.

Exercises

11.7.1   A sample survey was proposed to study the transportation habits of households located in a circular geographical region around the terminus of a commuter railroad. Two different types of sampling units were discussed. One type had sampling units shaped like pieces of pie radiating out from the central terminus. The second type of sampling grouped blocks of households into concentric circles surrounding the terminal. If the transportation characteristics of the households differ with increased distance from the terminus which sampling units would you recommend and why? Does your choice depend upon the use of a particular method of sampling?

11.7.2   In the manufacture of drugs, quality control requires daily estimation of some production characteristics. The daily production consists of 400 batches of 100 doses each. The batches are separate production runs produced throughout the day and there is reason to believe that much of the variability is among batches, especially among batches that are not produced near the same time of day. The three sampling plans given below are suggested. Cost considerations dictate a total sample size

---

2. Such a design would be called a stratified two stage sampling design.

of 40 doses.

(a) Divide the total day's production of 40,000 doses into eight hourly time periods and sample five doses from each hour's production by completely random sampling.

(b) View the 400 batches as sampling units and sample 20 batches and two doses from each batch.

(c) Sample 1 batch from the 20 produced in both the first and last hour of the day. Then sample three batches from the 60 produces in each of the other production hours of the day. Then from each batch select two doses for testing.

(d) Systematically select every 20th batch and then select two doses from the select batches.

Discuss each of the above sampling schemes separately. Be sure to discuss the potential strengths and weaknesses of each scheme in the light of any practical difficulties that could arise. For each sampling scheme list the questions you would like to have answered before recommending any sampling plan.

11.7.3   Imagine that the goal of a sociological study of a small city is the estimation of average family income. To do this the city is divided into $N = 400$ city "blocks". The blocks are carefully arranged so that each one has approximately 20 households in it. [For this exercise assume that all blocks contain exactly 20 households, that is all $M_i = 20$.] The sample design is a two stage design in which 40 blocks are selected and two households within each block. The sample estimate of $\hat{y}_{ms}$ is 9200; the "among blocks" sample variance is 3100 and the within block variance is 600 (1000's of squared dollars).

(a) Estimate the total income for the entire city.

(b) Assume that the estimates of variance given above are actually the true sampling variances and thereby calculate $Var(\hat{y}_{ms})$. That is, assume that

$$\frac{1}{N-1} \sum (Y_i - Y^*)^2 = 3100$$

and

$$\frac{1}{N} \sum_{i=1}^{N} S_i^2 = 600$$

(c) Calculate $Var(\hat{y}_{ms})$ if one household is selected from each of 80 blocks.

(d) Calculate $Var(\hat{y}_{ms})$ if four households are selected from each of 20 blocks.

(e) Suppose that it takes $10 in expenses to get an interviewer to each of the city blocks and $25 to locate and conduct each interview. Calculate the cost of each of the three sample allocations discussed in this exercise.

(f) Compare the relative costs of the three sample allocations with the relative variances. Discuss the results.

11.7.4   Give one word answers to the following questions

(a) In practice would you usually expect that putting "alikes" together in the same stratum would increase or decrease the sampling variance of the mean?

(b) Would you expect that putting "alikes" in the same primary would increase or decrease the variance of the mean?

(c) Would you expect that putting "unalikes" together in any sampling unit would increase or decrease the variance?

11.7.5   Given a sample of size four, how would you sample the four groups given in Exercise 11.3.2. Give reasons for your choice of sampling design and calculate the variance of your estimator of $\bar{Y}$. Compare this variance with the variance calculated in Exercise 11.3.2. Answer these same questions for the groups of Exercise 11.3.3.

11.8 Self-Weighting Designs and Data Bases

A self-weighting design is one in which unbiased estimation is associated with the equal weighting of each observation. In simple random sampling, the unbiased sample mean

$$\bar{y} = \frac{1}{n}(y_1 + \cdots + y_n) = \frac{1}{n}y_1 + \frac{1}{n}y_2 + \cdots + \frac{1}{n}y_n \quad (11.31)$$

weights each of the n observations by the same factor, $1/n$, and so the design is self-weighting. On the other hand, stratified sampling is not generally self-weighting. The stratified estimator, $\bar{y}_{st}$ (11.6), weights each observation in the $\underline{ith}$ stratum by $M_i/(Mm_i)$. This weight is the same for all units within the same stratum but differs from stratum to stratum. So stratified sampling is not generally self weighting.

However, stratified sampling can be made to be self-weighting by an appropriate allocation of the sample to strata. If sample units are allocated in proportion to the size of the strata, that is,

$$m_i = \frac{M_i}{M}m, \quad i = 1, 2, \ldots, N, \quad (11.32)$$

then,

$$\bar{y}_{st} = \sum_i \frac{M_i}{M} \frac{M}{m_i} \frac{1}{m} \sum_j y_{ij}$$
$$= \frac{\sum\sum y_{ij}}{m} \qquad\qquad (11.33)$$

and now each observation receives weight $1/m$. So in this special case stratified sampling is self-weighting.

Proportional allocation is sometimes also referred to as sampling with a *fixed* sampling fraction. The reason for this is that proportional allocation can be written as $m_i/M_i = m/M$, for all i. In this form, it is clear that the sampling fraction is the same in all strata.

In general, multistage designs are not self-weighting but can be made to be self-weighting in the same manner as stratified sampling.

The advantage of self-weighting designs is easy calculation and to a certain extent, easy understanding. With modern computers easy calculation is not always a big asset, but on the other hand, many analysts find unequal weighting difficult to understand. As a result, self-weighting designs are often a big help.

The availability of the computers has prompted a growing fascination with data bases. Unfortunately, documentation of these data bases often includes little or no description of the sampling process by which the data were obtained. As a result, analysts treat the data as if they *were* derived by a self-weighting design when in fact they were not! As we saw in Sections 10.2 and 10.3, using equal weights with unequal probability sampling biases the estimates. Consequently, it seems appropriate that data base engineers should either, make the details (weights) of the sampling design available to prevent misleading analysis, or, use a self-weighting design.

The use of self-weighting designs can be seriously hindered by nonresponse. Nonresponse is the failure to make measurements on units which have been randomly selected for inclusion in the sample. It occurs for many reasons, including failure of measuring devices and refusal to talk to interviewers. One important result is that nonresponse can destroy a self-weighting design by changing the sample allocation. Consequently, if nonresponse occurs it raises the question of how to present data to users. There appears to be two choices.

The first choice is to ignore the missing data and present the user with the weights appropriate for the *actual* realized sample allocation. This procedure implicitly assumes that the missing data are "like" the existing data, otherwise estimates relating to the target population will suffer from a selection bias. To illustrate, if a 50% response to a population survey is obtained, ignoring the missing data and basing estimates only on the existing data implies that the missing data are "like" the actual data. If they are not, then the estimates are biased.

The substitution of "typical" values for the missing observations is a second possible procedure. This preserves the self-weighting feature of the design but requires a similar assumption to the first procedure, namely that the missing observations can reasonably be derived by the available ones. Which of these two procedures is preferable does not seem to have a general answer, even if one overcomes the difficulty of defining and estimating "typical" values.

Finally, a word of warning about self-weighting designs. In the case of proportional allocation, the stratified estimator, (Expression 11.33), *looks like* the equally weighted mean, $\bar{y}$, used in simple random sampling. But these two estimators do not have the same variance. Even in the case of a self-weighting allocation, $\bar{y}_{st}$ requires use of the stratified variance formula; it is *not* correct to use the variance formula of the mean in simple random sampling, even though the algebraic form of the estimators is identical.

IMPORTANT NEW IDEAS

strata
cluster sampling
primaries, secondaries, etc.,
proportional allocation
nonresponse

stratified sampling
multistage sampling
optimum allocation
self-weighting designs

CHAPTER 12

AVERAGES AREN'T
EVERYTHING

Unequal probability sampling, multistage sampling, and stratified sampling were discussed in Chapters 10 and 11. Each of these schemes involves the method of sample selection and so must be specified *in advance,* as part of the sampling design. Consequently, we ask whether methods for reducing variance exist which do not involve the method of sample selection. If such methods do exist, they will have the very desirable operational feature that they can be utilized *after* the sample has been drawn. Since sample estimates are calculated after the sample has been drawn, any estimator with small variance would clearly have this desirable feature. In this chapter, we confine ourselves to simple random sampling and ask: Can estimators be found which have variances smaller than that of $\bar{y}$? The answer is yes, and as we shall see, the estimators of this type that are of most importance depend upon the availability of measurements over and above those of immediate interest.

In Section 12.1 some estimators other than $\bar{y}$ are discussed. The median is among those included because it is very popular, in spite of some weaknesses. Also, an estimator, which is a generalization of the median, is included - to expose the reader further to estimators which are insensitive to outlying observations. Finally, an estimator is described which successfully reduces variance by utilizing some specific information about the range of the population.

In Sections 12.3 and 12.4 ratio and regression estimators are discussed. These estimators are very important in applied sampling and are based on the availability of multiple measurements on the sample units. They are frequently better than $\bar{y}$ in simple random sampling.

A post-stratified estimator is discussed in Section 12.5. This estimator involves placing units into their appropriate strata *after* they have been drawn into the sample. It uses the same information as stratified sampling except that this information is used in the estimator rather than in the original sample selection procedure.

176

It is important to recognize that a smaller variance alone is not necessarily helpful. To illustrate, an estimator which is always equal to "1" has a smaller variance than $\bar{y}$, in fact it has *zero* variance, but clearly because of its severe bias, it is useless as an estimator of $\bar{Y}$! A more useful objective is to have estimators of $\bar{Y}$ which have smaller variance than $\bar{y}$ and which at the same time are unbiased or very close to unbiased.

12.1 Some Possibilities

12.1.1 Robust Estimators

Since the median is a popular estimator, we consider it first. Recall that the median is the middle value of an odd set of numbers and is the average of the two innermost values for an even set. So for samples of size 2, the median is the average of the two values. But this is exactly the same as the sample mean and so when $n = 2$ the mean and the median are identical, and as a result, will have exactly the same sampling distribution. Consequently, the comparison of the mean and median first becomes interesting when $n = 3$, then the sample median is no longer always equal to the sample mean. For samples of size 3 from the reduced ($N = 8$) taxpayer population, the sampling distributions of the mean, $\bar{y}$, and the median, $\hat{y}_{med}$, are given in Table 12.1 The estimator $\hat{y}_w$, which is also included in Table 12.1, is introduced later.

First notice that the expectation of $\hat{y}_{med}$ is equal to 90.21, so $\hat{y}_{med}$ is not an unbiased estimator of the population median, $Y_{med} = 92.00$. Nor is it unbiased for the population mean, $\bar{Y}_8 = 91.50$. These biases are not devastatingly large but unfortunately, $\hat{y}_{med}$ also has a bigger sampling variance than $\bar{y}$, 211.16 compared to 122.76; see Table 12.1. However, it is important to understand that this larger variance is *not* a general property of the median in comparison with the mean. In fact, even in this taxpayer example, if sampling is from the complete population rather than the reduced one, the comparison changes. Table 12.2 lists the sampling distributions of $\bar{y}$ and $\hat{y}_{med}$ for samples of size 3 from the complete population. Table 12.2 also details an estimator $\hat{y}_A$ which is introduced later.

In summary of Tables 12.1 and 12.2, the variance results for the mean and median are:

	$N = 8$	$N = 9$
$Var(\bar{y})$	122.76	405.32
$Var(\hat{y}_{med})$	211.26	314.51

TABLE 12.1

SAMPLING DISTRIBUTIONS OF $\bar{y}$, $\hat{y}_{med}$, $\hat{y}_w$

(3 taxpayers from 8)

Sample Number	Sampled Taxpayers	$\bar{y}$	$\hat{y}_{med}$	$\hat{y}_w$
1	1 2 3	66.67	68.00	67.33
2	1 2 4	75.33	72.00	73.67
3	1 2 5	74.00	72.00	73.00
4	1 2 6	78.00	72.00	75.00
5	1 2 7	82.67	72.00	77.33
6	1 2 8	87.33	72.00	79.67
7	1 3 4	74.00	68.00	71.00
8	1 3 5	72.67	68.00	70.33
9	1 3 6	76.67	68.00	72.00
10	1 3 7	81.33	68.00	74.67
11	1 3 8	86.00	68.00	77.00
12	1 4 5	81.33	90.00	85.67
13	1 4 6	85.33	94.00	89.67
14	1 4 7	90.00	94.00	92.00
15	1 4 8	94.67	94.00	94.33
16	1 5 6	84.00	90.00	87.00
17	1 5 7	88.67	90.00	89.33
18	1 5 8	93.33	90.00	91.67
19	1 6 7	92.67	102.00	97.33
20	1 6 8	97.33	102.00	99.67
21	1 7 8	102.00	116.00	109.00
22	2 3 4	78.00	72.00	75.00
23	2 3 5	76.67	72.00	74.33
24	2 3 6	80.67	72.00	76.33
25	2 3 7	85.33	72.00	78.67
26	2 3 8	90.00	72.00	81.00
27	2 4 5	85.33	90.00	87.67
28	2 4 6	89.33	94.00	91.67
29	2 4 7	94.00	94.00	94.00
30	2 4 8	98.67	94.00	96.33

TABLE 12.1 (continued)

SAMPLING DISTRIBUTIONS OF $\bar{y}$, $\hat{y}_{med}$, $\hat{y}_w$

(3 taxpayers from 8)

Sample Number	Sampled Taxpayers	$\bar{y}$	$\hat{y}_{med}$	$\hat{y}_w$
31	2 5 6	88.00	90.00	89.00
32	2 5 7	92.67	90.00	91.33
33	2 5 8	97.33	90.00	93.67
34	2 6 7	96.67	102.00	99.33
35	2 6 8	101.33	102.00	101.67
36	2 7 8	106.00	116.00	111.00
37	3 4 5	84.00	90.00	87.00
38	3 4 6	98.00	94.00	91.00
39	3 4 7	92.67	94.00	93.33
40	3 4 8	97.33	94.00	96.67
41	3 5 6	86.67	90.00	88.33
42	3 5 7	91.33	90.00	90.67
43	3 5 8	96.00	90.00	93.00
44	3 6 7	95.33	102.00	98.67
45	3 6 8	100.00	102.00	101.00
46	3 7 8	104.67	116.00	110.00
47	4 5 6	95.33	94.00	94.67
48	4 5 7	100.00	94.00	94.67
49	4 5 8	104.67	94.00	99.33
50	4 6 7	104.00	102.00	103.00
51	4 6 8	108.67	102.00	105.33
52	4 7 8	113.33	116.00	114.67
53	5 6 7	102.67	102.00	102.33
54	5 6 8	107.33	102.00	104.67
55	5 7 8	112.00	116.00	114.00
56	6 7 8	116.00	116.00	116.00
TOTAL		5124.00	5052.00	5088.00
EXPECTATION		91.50	90.21	90.87
VARIANCE		122.76	211.26	153.97

TABLE 12.2

SAMPLING DISTRIBUTION OF $\bar{y}$, $\hat{y}_{med}$, $\hat{y}_w$, $\hat{y}_A$

(3 taxpayers from 9)

Sample Number	Sampled Taxpayers	$\bar{y}$	$\hat{y}_{med}$	$\hat{y}_w$	$\hat{y}_A$
1	1 2 3	66.67	68.00	67.33	71.67
2	1 2 4	75.33	72.00	73.67	80.33
3	1 2 5	74.00	72.00	73.00	79.00
4	1 2 6	78.00	72.00	75.00	83.00
5	1 2 7	82.67	72.00	77.33	87.67
6	1 2 8	87.33	72.00	79.67	92.33
7	1 2 9	110.67	72.00	91.33	110.67
8	1 3 4	74.00	68.00	71.00	79.00
9	1 3 5	72.67	68.00	70.33	77.67
10	1 3 6	76.67	68.00	72.33	81.67
11	1 3 7	81.33	68.00	74.67	86.33
12	1 3 8	86.00	68.00	77.00	91.00
13	1 3 9	109.33	68.00	88.67	109.33
14	1 4 5	81.33	90.00	85.67	86.33
15	1 4 6	85.33	94.00	89.67	90.33
16	1 4 7	90.00	94.00	92.00	95.00
17	1 4 8	94.67	94.00	94.33	99.67
18	1 4 9	118.00	94.00	106.00	118.00
19	1 5 6	84.00	90.00	87.00	89.00
20	1 5 7	88.67	90.00	89.33	93.67
21	1 5 8	93.33	90.00	91.67	98.33
22	1 5 9	116.67	90.00	103.33	116.67
23	1 6 7	92.67	102.00	97.33	97.67
24	1 6 8	97.33	102.00	99.67	102.33
25	1 6 9	120.67	102.00	111.33	120.67
26	1 7 8	102.00	116.00	109.00	107.00
27	1 7 9	125.33	116.00	120.67	125.33
28	1 8 9	130.00	130.00	130.00	130.00
29	2 3 4	78.00	72.00	75.00	78.00
30	2 3 5	76.67	72.00	74.33	76.67
31	2 3 6	80.67	72.00	76.33	80.67
32	2 3 7	85.33	72.00	78.67	85.33
33	2 3 8	90.00	72.00	81.00	90.00
34	2 3 9	113.33	72.00	92.67	108.33
35	2 4 5	85.33	90.00	87.67	85.33
36	2 4 6	89.33	94.00	91.67	89.33
37	2 4 7	94.00	94.00	94.00	94.00
38	2 4 8	98.67	94.00	96.33	98.67
39	2 4 9	122.00	94.00	108.00	117.00
40	2 5 6	88.00	90.00	89.00	88.00

Sample Number	Sampled Taxpayers	$\bar{y}$	$\hat{y}_{med}$	$\hat{y}_w$	$\hat{y}_A$
41	2 5 7	92.67	90.00	91.33	92.67
42	2 5 8	97.33	90.00	93.67	97.33
43	2 5 9	120.67	90.00	105.33	115.67
44	2 6 7	96.67	102.00	99.33	96.67
45	2 6 8	101.33	102.00	101.67	101.33
46	2 6 9	124.67	102.00	113.33	119.67
47	2 7 8	106.00	116.00	111.00	106.00
48	2 7 9	129.33	116.00	122.67	124.33
49	2 8 9	134.00	130.00	132.00	129.00
50	3 4 5	84.00	90.00	87.00	84.00
51	3 4 6	88.00	94.00	91.00	88.00
52	3 4 7	92.67	94.00	93.33	92.67
53	3 4 8	97.33	94.00	95.67	97.33
54	3 4 9	120.67	94.00	107.33	115.67
55	3 5 6	86.67	90.00	88.33	86.67
56	3 5 7	91.33	90.00	90.67	91.33
57	3 5 8	96.00	90.00	93.00	96.00
58	3 5 9	119.33	90.00	104.67	114.33
59	3 6 7	95.33	102.00	98.66	95.33
60	3 6 8	100.00	102.00	101.00	100.00
61	3 6 9	123.33	102.00	112.67	118.33
62	3 7 8	104.67	116.00	110.33	104.67
63	3 7 9	128.00	116.00	122.00	123.00
64	3 8 9	132.67	130.00	131.33	127.67
65	4 5 6	95.33	94.00	94.67	95.33
66	4 5 7	100.00	94.00	97.00	100.00
67	4 5 8	104.67	94.00	99.33	104.67
68	4 5 9	128.00	94.00	111.00	123.00
69	4 6 7	104.00	102.00	103.00	104.00
70	4 6 8	108.67	102.00	105.33	108.67
71	4 6 9	132.00	102.00	117.00	127.00
72	4 7 8	113.33	116.00	114.67	113.33
73	4 7 9	136.67	116.00	126.33	131.67
74	4 8 9	141.33	130.00	135.67	136.33
75	5 6 7	102.67	102.00	102.33	102.67
76	5 6 8	107.33	102.00	104.67	107.33
77	5 6 9	130.67	102.00	116.33	125.67
78	5 7 8	112.00	116.00	114.00	112.00
79	5 7 9	135.33	116.00	125.67	130.33
80	5 8 9	140.00	130.00	135.00	135.00
81	6 7 8	116.00	116.00	116.00	116.00
82	6 7 9	139.33	116.00	127.67	134.33
83	6 8 9	144.00	130.00	137.00	139.00
84	7 8 9	148.67	130.00	139.33	143.67
TOTAL		8698.67	8026.00	8362.33	8698.67
MEAN		103.56	95.55	99.55	103.56
VARIANCE		405.32	314.51	318.63	301.12

These calculated variances lead to the question, "Why does the median do better than $\bar{y}$ when sampling from the complete population and worse when sampling from the reduced population?" The answer is that the ninth taxpayer has a very large income, $y_9 = 200$, which when selected affects the sample mean $\bar{y}$ very heavily. On the other hand, the median is not influenced as much by the size of y_9. As a result of this property, in real studies of personal income, the existence of a few individuals with large incomes frequently results in a decision to use the median.

The lack of sensitivity of the median to large and small observations, in comparison with $\bar{y}$, can be seen by writing the two estimators as follows:

$$\bar{y} = \frac{1}{3} \times \left\{ \begin{array}{c} smallest \\ observation \end{array} \right\} + \frac{1}{3} \times \left\{ \begin{array}{c} middle \\ observation \end{array} \right\} + \frac{1}{3} \times \left\{ \begin{array}{c} largest \\ observation \end{array} \right\}$$

$$\hat{y}_{med} = 0 \times \left\{ \begin{array}{c} smallest \\ observation \end{array} \right\} + 1 \times \left\{ \begin{array}{c} middle \\ observation \end{array} \right\} + 0 \times \left\{ \begin{array}{c} largest \\ observation \end{array} \right\}. \quad (12.1)$$

In this form, it is clear that $\bar{y}$ gives equal weight to each observation while $\hat{y}_{med}$ gives no weight at all to the largest and smallest measurements. This leads us to ask if an estimator that is better than either $\bar{y}$ or $\hat{y}_{med}$ can be found by giving the largest and smallest observations weights which are greater than zero but less than 1/3. Here, "better" means having a smaller bias than $\hat{y}_{med}$ while at the same time retaining the property of $\hat{y}_{med}$ that its variance is not unduly inflated by the large value of y_9.

For example, consider the estimator $\hat{y}_w$ which gives the three ordered sample observations weights $\{1/6, 4/6, 1/6\}$, instead of the weights $\{0, 1, 0\}$ used by $\hat{y}_{med}$ and the weights $\{1/3, 1/3, 1/3\}$ used by $\bar{y}$. This estimator gives more weight to the largest and smallest observations than the median, but not as much as the equally weighted mean $\bar{y}$. So in sample number 1 (Table 12.1) with taxpayers 1, 2, and 3,

$$\hat{y}_w = \frac{1}{6} (60) + \frac{4}{6} (68) + \frac{1}{6} (72) = 67.3. \quad (12.2)$$

The sampling distribution of $\hat{y}_w$ appears in both Table 12.1 (3 taxpayers from 8) and Table 12.2 (3 taxpayers from 9). The relevant results from the two tables are summarized in Table 12.3.

From Table 12.3 we see that in the reduced population $\bar{y}$ is clearly best. It is unbiased and has the smallest variance of the three

TABLE 12.3

BIAS AND VARIANCE OF ROBUST ESTIMATORS

	N = 8		N = 9	
Estimator	Bias for $\overline{Y}$ = 91.50	Variance	Bias for $\overline{Y}$ = 103.56	Variance
$\overline{y}$	0	122.76	0	405.32
$\hat{y}_{med}$	-1.29	211.26	-8.01	314.51
$\hat{y}_w$	-0.63	153.97	-4.01	318.63

estimators. Furthermore, $\hat{y}_w$ has both a smaller bias and a smaller variance than $\hat{y}_{med}$, so $\hat{y}_w$ would appear to be the second choice for sampling from the reduced population.

The situation is *not* the same when sampling from the complete population. The price for the unbiasedness of $\overline{y}$ is a variance that is larger than that of either $\hat{y}_{med}$ or $\hat{y}_w$. The choice between $\hat{y}_{med}$ and $\hat{y}_w$ is not as clear as it was in the reduced population, but if the choice is based on mean square error, $\hat{y}_w$ would again be selected over $\hat{y}_{med}$.

The reason for the downfall of $\overline{y}$ when sampling from the complete population lies in the fact that T_9 has an income that is very much larger than the other taxpayers'. Consequently, the estimators $\hat{y}_{med}$ and $\hat{y}_w$ which give T_9 smaller weight than $\overline{y}$ are less affected by his inclusion. Estimators with this property are called "robust" (against extreme observations). Such estimators are not discussed further in this book, but have been introduced to add to the readers understanding of estimators.

12.1.2 Auxiliary Information

The estimators discussed in Section 12.1.1 did not use any information in addition to the sample y measurements. The knowledge of additional information often leads to more efficient estimators. To illustrate, suppose it is *known* that T_1 has the smallest income and T_9 the largest. To take advantage of this knowledge, add 15 to $y_1 = 60$ whenever taxpayer 1 appears in a sample and subtract 15 from $y_9 = 200$

whenever taxpayer 9 appears. The other observations, $y_2, y_3, y_4, y_5, y_6, y_7$, and y_8 are unchanged. With these modifications, calculate the equally weighted mean for each sample. These calculations have all been made and are listed in Table 12.2. However, to distinguish this case from $\bar{y}$ based on the unadjusted measurements of T_1 and T_9, the new estimator has been labeled as $\hat{y}_A$. As an example, if sample 1 is selected,

$$\hat{y}_A = (75+68+72)/3 = 71.67. \tag{12.3}$$

The hope is that these changes, which clearly reduce the range of the sampled numbers, also reduce the variance of the sampling distribution.

From Table 12.2, it can be seen that the estimator $\hat{y}_A$ is unbiased and has smaller variance than $\bar{y}$. The adjustment made possible by knowledge about the largest and smallest taxpayers has indeed resulted in a more desirable estimator. But unfortunately, information of this type is not normally available in real problems; however, in the next section we shall discuss a type of auxiliary information which frequently *is* available in practice.

Before finishing this introductory section on estimators, it is worth noticing that we were led implicitly by analogy in Chapter 5 to the sample mean $\bar{y}$ as an estimator of the population mean $\bar{Y}$, that is, the sample mean is the same calculation on the sample numbers as the population mean is on the entire set of population numbers. In this case, the analogy principle worked well because the sample mean is frequently a good estimator. Unfortunately, the analogy principle does not always work so well. In fact in general, it guarantees neither unbiasedness nor good variance properties. The characteristics of "analogy" estimators are not discussed further in this book; the point has been brought up only to serve as a warning that this procedure is not always a good one to follow.

Exercises

12.1.1   Randomly select three of the numbered samples in Table 12.1 and verify the calculations given there for $\bar{y}$, $\hat{y}_{med}$, and $\hat{y}_w$.

12.1.2   Calculate the numerical value of the bias of the estimators $\hat{y}_{med}$ and $\hat{y}_w$ in the sampling distributions given in Table 12.1. Make the same calculation for the sampling distributions in Table 12.2. Compare and discuss these biases.

12.1.3   Calculate the mean square errors for each of the six cases (three estimators and two populations) included in Table 12.3. Compare and discuss these mean square errors.

12.1.4   Randomly select three of the numbered samples in Table 12.2 and verify the calculations given there for $\hat{y}_A$.

12.1.5   Calculate the mean square error of $\hat{y}_A$ (Table 12.1) and compare it with the six mean square errors calculated in Exercise 12.1.3.

12.2 Related Measurements

In the taxpayer example, *both* actual income (y) and reported income (x) are discussed. So far, most interest has focused on actual income which is presumed to require a special audit to determine it. On the other hand, reported income x is known for *every* taxpayer without an audit,[1] and as a result the true mean, $\overline{X}$, and total, X, of the reported incomes are also known *exactly*. The question then arises, "Is it possible that this information can be utilized to reduce variance just as additional information was used in the estimator $\hat{y}_A$ to reduce variance?" As we shall see, the information on x is very valuable indeed.

For the taxpayer population, the relationship between actual and reported income is shown in Figure 12.1. The large taxpayer, who was excluded from the sampling process in Chapter 5, is included here because the estimation schemes that are about to be developed can accommodate him very nicely.

As we shall see, *ratio and regression estimators* actually utilize the strength of the linear relationship between y and x, that is, how close the points (y_i, x_i) are to falling on a straight line. This strength is measured by the *linear* correlation coefficient,

$$\rho = \frac{\sum (y_i - \overline{Y})(x_i - \overline{X})}{\sqrt{\sum (y_i - \overline{Y})^2}\sqrt{\sum (x_i - \overline{X})^2}} \tag{12.4}$$

where ρ is the Greek letter "rho." As in our earlier uses of Greek

---

1. The assumption that every person required to file a tax return actually does so is needed at this point. Such an assumption may amuse some readers, but making it will not mislead readers as to the nature and utility of the sampling methods.

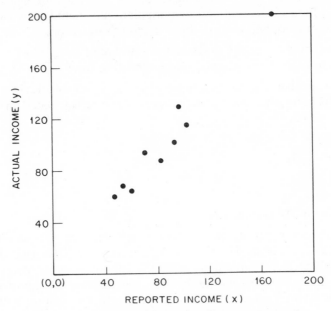

Figure 12.1. Relationship of reported and actual
incomes ($\rho = 0.98$)

letters, ρ is the most common statistical notation for this correlation
coefficient.

The coefficient ρ can take on values between $+1.0$ and -1.0. It
is $+1.0$ when *all* the points are *exactly* on a straight line with positive
slope and is -1.0 when all the points fall exactly on a negatively sloped
line; these are the cases of perfect correlation. When $\rho = 0$ there is no
linear relationship. Some examples are shown in Figure 12.2. In the
population of nine taxpayers, $\rho = 0.98$, which is a very high linear
correlation. This can be confirmed intuitively by simply placing a ruler
on Figure 12.1 and observing that the nine points are close to a straight
line.

It is important to emphasize that ρ measures *linear* correlation.
For the five points $y,x=(4, -2)$, $(1, -1)(0,0)(1,1)(4,2)\}$, $\rho = 0$ yet
they are all exactly on the curve $y = x^2$. Thus, these points have zero
linear correlation but at the same time have a perfect curvilinear relationship. Although we shall not make much of a point of it in this
book, it clearly is a mistake to calculate ρ without examining the data
for *non*linear relationships.

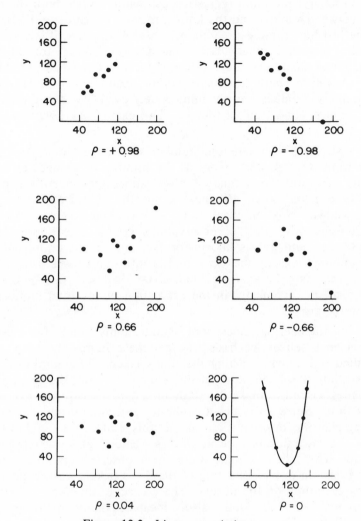

Figure 12.2. Linear correlations.

In summary, ratio and regression estimators utilize both the relationship between y and x and the known mean $\bar{X}$ to estimate $\bar{Y}$. Before discussing just how these estimators use this information, it is appropriate to ask, "Is such auxiliary data realistically available?" Indeed it is. Even in a real taxpayer study, the x_i's for most persons should be readily available. There are many similar examples; even in cases where auxiliary x values are not immediately available they can sometimes be *created,* almost out of thin air. Two examples are very enlightening.

In a study to estimate the number of oranges in a grove, an observer went through and "guessed" the number of oranges on *every* tree. This guess was the auxiliary x_i observation, and since it was made on every tree in the grove, the mean $\bar{X}$ and the total X of the guesses were also known. Next an exact count, y_i, was obtained for a *sample* of trees. Consequently, pairs of observations (y_i, x_i) were known for a sample of trees, and x_i was known for each tree in the entire population. In this way, exactly the same kind of information was made available as in the taxpayer example, and as we shall see, this "guess" or "eye estimate" is very useful in the estimation of the total number of oranges in the grove.

In an engineering stockroom, some items were available to employees on a selfservice basis. To get these items, the employees simply filled out a ticket, listing the items taken. The administration wished to study the dollar value of items removed from the stockroom. Further, they wished to make the study on a daily basis for an extended period of time. Since every ticket could not be economically evaluated, a sampling plan was in order. To conduct this study, at the end of each day all of the day's tickets were *weighed* as a batch. This was cheap and easy to do. These daily weights were the x observations. Then on a sampling basis, an exact evaluation of the day's tickets was obtained by examining each item on the tickets. These exact evaluations were the y measurements. In this study also, cheap auxiliary measurements greatly aided the efficiency of the sampling study.

Exercises

12.2.1   Randomly select three of the numbered samples given in Table 12.2 and verify the calculations given there for $\bar{y}$, $\hat{y}_{med}$, $\hat{y}_w$, and $\hat{y}_A$.

12.2.2   Verify that the linear correlation, ρ, for the five points $\{y,x\}:(4, -2)$, $(1, -1)$, $(0,0)$, $(1,1)$, $(4,2)$ is equal to zero. Plot these five points on a graph and verify that they all lie exactly on the quadratic $y = x^2$.

12.2.3   Verify numerically that the correlation between actual and reported incomes in the taxpayer population is 0.98, as given in Figure 12.1. These points are also plotted in Figure 12.2.
Some hand calculators have the computation of the correlation coefficient preprogrammed. If you have one of these the calculation of a correlation is trivial. If you do not, use the formula

$$\rho = \frac{\sum y_i x_i - n\overline{y}\overline{x}}{\sqrt{(\sum x_i^2 - n\overline{x}^2)(\sum y_i^2 - n\overline{y}^2)}}$$

This formula requires computation of means, sums of squares and cross products which is easily done on many calculators.

12.2.4   In Section 12.3 two studies were described in which auxiliary variates were "created." Make a list of all the potential areas of sampling application in which you can imagine an auxiliary being usefully "created."

12.2.5   In Figure 12.2 the y values in the plot with $\rho = -0.98$ are obtained from the y values of the $\rho = 0.98$ plot by changing the y values to $200 - y$. Use the formula for ρ in Section 12.2 to show *algebraically* that given $\rho = 0.98$ the correlation for the derived data set *must* be -0.98.

12.2.6   Numerically verify the correlations of 0.66, 0.04 and 0 in the remaining plots of Figure 12.2. The (y,x) values of the plotted points are:

$\rho = 0.66$	$\rho = -0.66$	$\rho = 0.04$	$\rho = 0$
100 50	100 50	100 50	180 80
85 85	115 85	85 85	100 88
60 105	140 105	60 105	60 100
120 110	80 110	120 110	20 120
110 120	90 120	110 120	60 140
75 135	125 135	75 135	100 152
105 145	95 145	105 145	180 160
125 150	75 150	125 150	
185 200	15 200	85 200	

12.2.7   If in the handling of the data set associated with the correlation of 0.66 in Figure 12.2 and Exercise 12.2.6, the "1" in the hundreds place of the number 185 is accidentally dropped the "erroneous" data set is then identical to the set of data in the plot below in Figure 12.2. The correlation between y and x for this set of data, which is also given in Exercise 12.2.6, is 0.04. From this one can make the observation that the single point, (185,200) or (85,200), has an enormous effect on the calculated correlation coefficient. Do you think that this property of extreme sensitivity of the correlation coefficient to this single is of any applied relevance? Discuss your answer.

12.2.8   In Figure 12.2, the seven points plotted in the graph with $\rho = 0$ lie on the parabola $y = 20 + 0.1(x - 120)^2$. Since the relationship between y and x is *quadratic* and we know that ρ measures *linear* correlation, the fact that $\rho = 0$ might have been anticipated. This exercise is intended to illustrate the dependency of ρ on scale. While the example is extreme, this feature is important in the interpretation of calculated correlation coefficients.

   i.   Transform the seven (y,x) points referred to above to the seven (y,z) points by using the relationship $z = (x - 120)^2$.

   ii.   Plot the seven points (y,z) in the same manner as Figure 12.2.

   iii.   Calculate the correlation between the two variables y and z.

   iv.   Discuss your findings.

12.3 Ratio Estimation

The estimator,

$$\hat{y}_r = \frac{\bar{y}}{\bar{x}} \, \bar{X}, \tag{12.5}$$

is the most commonly used ratio estimator of the population mean, $\bar{Y}$. In $\hat{y}_r$, $\bar{y}$ and $\bar{x}$ are sample means and $\bar{X}$ is the population mean. Since $\bar{X}$ is constant, the variability in $\hat{y}_r$ comes from the variability in the ratio of sample means, $\bar{y}/\bar{x}$. The hope is that basing the variability in $\hat{y}_r$ on

$\bar{y}/\bar{x}$ will result in an estimator with a smaller variance than the mean $\bar{y}$ alone.

Graphically, in Figure 12.3, the estimator $\hat{y}_r$ is formed by evaluating the straight line through the origin, $y = (\bar{y}/\bar{x})x$, at the known point, $\bar{X}$. The hope is that the slope of the line $\bar{y}/\bar{x}$, will have a small variance.

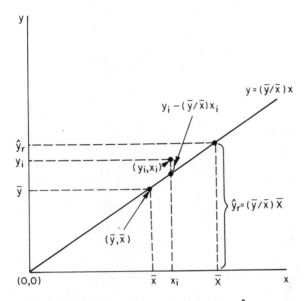

Figure 12.3. The ratio estimator, $\hat{y}_r$.

Since the sample data are utilized only in the form of a ratio, the estimator $\hat{y}_r$ is called a ratio estimator. This name is sometimes confusing. The estimator $\hat{y}_r$ does not estimate a ratio, it estimates the mean $\bar{Y}$, but it does it by use of a ratio.

To illustrate the calculation of $\hat{y}_r$, if sample number 1 (Table 12.4) with taxpayers 1, 2, and 3 (Table 3.2) is selected, then,

$$\hat{y}_r = \frac{66.67}{57.33} (92.78) = 107.88.$$

The complete sampling distribution of $\hat{y}_r$, for samples of three taxpayers from nine, is given in Table 12.4. The estimators $\hat{y}_b$ and $\hat{y}_B$ which are also in Table 12.4 are regression estimators and are discussed in the next section.

TABLE 12.4

SAMPLING DISTRIBUTION OF $\bar{y}, \hat{y}_r, \hat{y}_b, \hat{y}_B$

(3 taxpayers from 9)

Sample Number	Taxpayers Sampled	$\bar{y}$	$\hat{y}_r$	$\hat{y}_b$	$\hat{y}_B$
1	1 2 3	66.67	107.88	81.13	105.51
2	1 2 4	75.33	115.21	115.65	110.53
3	1 2 5	74.00	105.09	92.52	104.08
4	1 2 6	78.00	105.39	96.83	104.43
5	1 2 7	82.67	105.55	99.89	104.71
6	1 2 8	87.33	112.54	110.81	110.11
7	1 2 9	110.67	109.62	109.69	109.69
8	1 3 4	74.00	107.27	109.47	105.54
9	1 3 5	72.67	98.18	91.07	99.09
10	1 3 6	76.67	98.79	94.64	99.44
11	1 3 7	81.33	99.29	97.01	99.72
12	1 3 8	86.00	105.91	107.25	105.12
13	1 3 9	109.33	104.57	104.46	104.71
14	1 4 5	81.33	104.80	98.38	104.11
15	1 4 6	85.33	105.09	100.01	104.45
16	1 4 7	90.00	105.25	101.90	104.74
17	1 4 8	94.67	111.65	111.05	110.13
18	1 4 9	118.00	109.11	109.64	109.72
19	1 5 6	84.00	97.42	94.40	98.00
20	1 5 7	88.67	97.93	96.41	98.29
21	1 5 8	93.33	103.91	103.79	103.68
22	1 5 9	116.67	103.08	102.64	103.27
23	1 6 7	92.67	98.44	97.48	98.63
24	1 6 8	97.33	104.20	103.82	104.03
25	1 6 9	120.67	103.34	103.02	103.62
26	1 7 8	102.00	104.38	104.17	104.31
27	1 7 9	125.33	103.51	103.42	103.90
28	1 8 9	130.00	108.01	108.86	109.30
29	2 3 4	78.00	109.65	107.46	107.35
30	2 3 5	76.67	100.66	89.94	100.90
31	2 3 6	80.67	101.14	95.00	101.25
32	2 3 7	85.33	101.50	98.12	101.53
33	2 3 8	90.00	107.97	108.15	106.93
34	2 3 9	113.33	106.21	106.29	106.51
35	2 4 5	85.33	106.99	96.01	105.91
36	2 4 6	89.33	107.17	99.69	106.26
37	2 4 7	94.00	107.23	102.75	106.54
38	2 4 8	98.67	113.48	111.66	111.94
39	2 4 9	122.00	110.61	111.74	111.53
40	2 5 6	88.00	99.57	94.93	99.81

Sample Number	Taxpayers Sampled	$\bar{y}$	$\hat{y}_r$	$\hat{y}_b$	$\hat{y}_B$
41	2 5 7	92.67	99.97	97.84	100.10
42	2 5 8	97.33	105.82	104.90	105.49
43	2 5 9	120.67	104.63	104.77	105.08
44	2 6 7	96.67	100.39	99.28	100.44
45	2 6 8	101.33	106.03	105.26	105.84
46	2 6 9	124.67	104.83	105.35	105.43
47	2 7 8	106.00	106.13	106.10	106.12
48	2 7 9	129.33	104.95	106.03	105.71
49	2 8 9	134.00	109.37	111.53	111.11
50	3 4 5	84.00	100.78	96.96	100.93
51	3 4 6	88.00	101.21	98.63	101.27
52	3 4 7	92.67	101.54	100.20	101.56
53	3 4 8	97.33	107.50	108.92	106.95
54	3 4 9	120.67	105.95	105.79	106.54
55	3 5 6	86.67	94.23	94.00	94.83
56	3 5 7	91.33	94.85	94.92	95.11
57	3 5 8	96.00	100.45	101.73	100.51
58	3 5 9	119.33	100.35	97.69	100.09
59	3 6 7	95.33	95.45	95.45	95.46
60	3 6 8	100.00	100.85	107.01	100.85
61	3 6 9	123.33	100.67	97.71	100.44
62	3 7 8	104.67	101.15	100.77	101.14
63	3 7 9	128.00	100.93	97.68	100.72
64	3 8 9	132.67	105.20	103.58	106.12
65	4 5 6	95.33	99.75	96.54	99.84
66	4 5 7	100.00	100.12	100.07	100.12
67	4 5 8	104.67	105.55	105.54	105.52
68	4 5 9	128.00	104.48	104.00	105.11
69	4 6 7	104.00	100.51	102.16	100.47
70	4 6 8	108.67	105.75	106.30	105.87
71	4 6 9	132.00	104.67	104.70	105.45
72	4 7 8	113.33	105.83	108.03	106.15
73	4 7 9	136.67	104.79	105.61	105.14
74	4 8 9	141.33	108.97	111.81	111.13
75	5 6 7	102.67	94.62	93.35	94.02
76	5 6 8	107.33	99.58	92.89	99.42
77	5 6 9	130.67	99.64	93.14	99.00
78	5 7 8	112.00	99.91	95.02	99.70
79	5 7 9	135.33	99.91	92.49	99.29
80	5 8 9	140.00	103.91	100.03	104.68
81	6 7 8	116.00	100.27	91.35	100.05
82	6 7 9	139.33	100.21	91.68	99.63
83	6 8 9	144.00	104.10	100.48	105.03
84	7 8 9	148.67	104.23	101.31	105.31
TOTAL		8698.67	8707.67	8503.84	8698.67
EXPECTATION		103.56	103.66	101.24	103.56
VARIANCE		405.32	18.04	39.90	16.00

First observe in Table 12.4 that $\hat{y}_r$ is biased as an estimator of $\overline{Y}$ by an amount $103.66 - 103.56 = 0.10$. This bias is a very small fraction of the standard error, $\sqrt{18.04} = 4.25$, so that almost all of the inaccuracy in $\hat{y}_r$ comes from variance and not from bias. This is a general characteristic of the estimator $\hat{y}_r$. It can be proven mathematically that $\hat{y}_r$ is biased and that the bias is usually negligible, even for moderately sized samples.

But the most important feature about $\hat{y}_r$ is that its variance, 18.04, is very much smaller than the variance of $\overline{y}$, 405.32; in fact, the ratio of these variances is about 1/20. This dramatic reduction in variance is directly related to the very high correlation between y and x in the population of taxpayers. While in practice, gains are not usually so spectacular, the use of ratio estimators often reduces variance by 1/2 to 3/4 over estimators which do not utilize the auxiliary data.

In this example, ratio estimation reduces variance more effectively than the special treatment of the big taxpayer which was discussed in Chapters 5 and 11. Recall that the large taxpayer was included in all samples while the rest of each sample was selected randomly from the remaining taxpayers. This special treatment resulted in a sampling variance of 174.65, see Table 11.2, which compares with $Var(\hat{y}_r) = 18.04$. The estimator $\hat{y}_r$ does well because the ratio $\overline{y}/\overline{x}$ is very stable from one sample to the next.

The *sample* variance of $\hat{y}_r$ can be calculated approximately by the mathematically derived formula,

$$var(\hat{y}_r) = \frac{1}{n(n-1)} \sum_{i=1}^{n} (y_i - \frac{\overline{y}}{\overline{x}} x_i)^2. \qquad (12.6)$$

From this formula, it can be seen that the linear relationship between y and x should pass through the origin; otherwise the differences $y_i - (\overline{y}/\overline{x})x_i$ (and hence $var(\hat{y}_r)$) will be inflated. To see this, look at Figures 12.1 and 12.3 and consider the effect of data points which are not close to a line through the origin. For such points the deviation $y_i - (\overline{y}/\overline{x})x_i$ is large.

When the linear relationship describing the (y_i, x_i) points does not pass near the origin the bias in $\hat{y}_r$ is increased as well as the variance. To illustrate, suppose that 30 is added to each y observation in the taxpayer population. The effect of this addition is that the relationship between y and x does not pass close to the origin. It also increases the true mean from 103.56 to 133.56, but does not change the variance of $\overline{y}$ in simple random sampling. So as before,

$$V(\overline{y}) = (1 - \frac{n}{N}) \frac{S^2}{n} = 405.32. \qquad (12.7)$$

As usual $\bar{y}$ is unbiased. Of course, $\bar{y}$ now estimates the new mean of $\bar{Y} = 133.56$.

The estimator $\hat{y}_r$ has not reacted as nicely as $\bar{y}$ to the addition of 30 to the y values. The expectation of $\hat{y}_r$ has changed from 103.66 to 134.82, see Table 12.5, so that the bias has increased from $103.66 - 103.56 = 0.10$ to $134.82 - 133.56 = 1.26$, a twelvefold increase! Also, since the addition of 30 does not affect the variance of $\bar{y}$ in simple random sampling, it might have been (erroneously) expected that this simple addition would have *no* increase in the variance of $\hat{y}_r$. However, observe (Table 12.5) that $Var(\hat{y}_r)$ has increased from 18.04 to 65.11, a factor of 3-1/2!

Fortunately, the bias in $\hat{y}_r$ can be shown to decrease as the sample size increases, regardless of whether the relationship passes through the origin or not. Nevertheless, if the intercept cutoff by the line relating y and x gets bigger so do both the bias and variance. So, if the sample size is small, some care must be taken in the use of $\hat{y}_r$. In practice, small samples occur most frequently within strata, so it is within strata that problems with the application of $\hat{y}_r$ are most likely to occur.

Exercises

12.3.1   Select three of the numbered samples from Table 12.4 and verify the calculations given there for $\bar{y}$, and $\hat{y}_r$.

12.3.2   Consider the following population of $N=5$ pairs (y_i, x_i); (2,1) (1,2), (3,3), (6,5), (7,9). List all possible samples of size $n=2$ that can be drawn from this population by sampling with equal probability without replacement.

12.3.3   For each of the samples listed in Exercise 12.3.2 calculate the value of the estimator $\bar{r} = (1/n) \sum (y_i/x_i)$.

12.3.4   Is the estimator $\bar{r}$ in Exercise 12.3.3 an unbiased estimator of the ratio $\bar{Y}/\bar{X}$? If not, calculate the bias.

12.3.5   Is the estimator $\bar{r}\bar{X}$, where $\bar{r} = (1/n) \sum (y_i/x_i)$ and $\bar{X}$ is the mean of the five x_i values, an unbiased estimator of $\bar{Y}$? If not, calculate the bias *and* the mean square error.

12.3.6   For each of the samples in Exercise 12.3.2 calculate the estimator

$$\hat{y}_{HR} = \bar{r}\bar{X} + \frac{N-1}{N} \frac{n}{n-1} (\bar{y} - \bar{r}x).$$

TABLE 12.5

SAMPLING DISTRIBUTIONS OF $\bar{y}$ AND $\hat{y}_r$

Effect of Relationship Not Passing Through Origin

Sample Number	Sampled Taxpayers	Small Intercept		Large Intercept	
		$\bar{y}$	$\hat{y}_r$	$\bar{y}$	$\hat{y}_r$
1	1 2 3	66.67	107.88	96.67	156.42
2	1 2 4	75.33	115.21	105.33	161.08
3	1 2 5	74.00	105.09	104.00	147.68
4	1 2 6	78.00	105.39	108.00	145.92
5	1 2 7	82.67	105.55	112.67	143.84
6	1 2 8	87.33	112.54	117.33	151.19
7	1 2 9	110.67	109.62	140.67	139.33
8	1 3 4	74.00	107.27	104.00	150.76
9	1 3 5	72.67	98.18	102.67	138.71
10	1 3 6	76.67	98.79	105.67	137.44
11	1 3 7	81.33	99.29	111.33	135.91
12	1 3 8	86.00	105.91	116.00	142.86
13	1 3 9	109.33	104.57	139.33	133.26
14	1 4 5	81.33	104.80	111.33	143.46
15	1 4 6	85.33	105.09	115.33	142.04
16	1 4 7	90.00	105.25	120.00	140.33
17	1 4 8	94.67	111.65	124.67	147.02
18	1 4 9	118.00	109.11	148.00	136.85
19	1 5 6	84.00	97.42	114.00	132.20
20	1 5 7	88.67	97.93	118.67	131.06
21	1 5 8	93.33	103.91	123.33	137.31
22	1 5 9	116.67	103.08	146.67	129.59
23	1 6 7	92.67	98.44	122.67	130.31
24	1 6 8	97.33	104.20	127.33	136.31
25	1 6 9	120.67	103.34	150.67	129.03
26	1 7 8	102.00	104.38	132.00	135.07
27	1 7 9	125.33	103.51	155.33	128.29
28	1 8 9	130.00	108.01	160.00	132.93
29	2 3 4	78.00	109.65	108.00	151.81
30	2 3 5	76.67	100.66	106.67	140.04
31	2 3 6	80.67	101.14	110.67	138.74
32	2 3 7	85.33	101.50	115.33	137.18
33	2 3 8	90.00	107.97	120.00	143.96
34	2 3 9	113.33	106.21	143.33	134.32
35	2 4 5	85.33	106.99	115.33	144.59
36	2 4 6	89.33	107.17	119.33	143.16
37	2 4 7	94.00	107.23	124.00	141.44
38	2 4 8	98.67	113.48	128.67	147.98
39	2 4 9	122.00	110.61	152.00	137.80
40	2 5 6	88.00	99.57	118.00	133.50

Sample Number	Sampled Taxpayers	Small Intercept		Large Intercept	
		$\bar{y}$	$\hat{y}_r$	$\bar{y}$	$\hat{y}_r$
41	2 5 7	92.67	99.97	122.67	132.33
42	2 5 8	97.33	105.82	127.33	138.44
43	2 5 9	120.67	104.63	150.67	130.64
44	2 6 7	96.67	100.39	126.67	131.55
45	2 6 8	101.33	106.03	131.33	137.42
46	2 6 9	124.67	104.83	154.67	130.06
47	2 7 8	106.00	106.13	136.00	136.16
48	2 7 9	129.33	104.95	159.33	129.29
49	2 8 9	134.00	109.37	164.00	133.86
50	3 4 5	84.00	100.78	114.00	136.77
51	3 4 6	88.00	101.21	118.00	135.72
52	3 4 7	92.67	101.54	122.67	134.42
53	3 4 8	97.33	107.50	127.33	140.64
54	3 4 9	120.67	105.95	150.67	132.29
55	3 5 6	86.67	94.23	116.67	126.84
56	3 5 7	91.33	94.85	121.33	126.01
57	3 5 8	96.00	100.45	126.00	131.84
58	3 5 9	119.33	100.35	149.33	125.57
59	3 6 7	95.33	95.45	125.33	125.48
60	3 6 8	100.00	100.85	130.00	131.10
61	3 6 9	123.33	100.67	153.33	125.15
62	3 7 8	104.67	101.15	134.67	130.15
63	3 7 9	128.00	100.93	158.00	124.58
64	3 8 9	132.67	105.20	162.67	128.99
65	4 5 6	95.33	99.75	125.33	131.14
66	4 5 7	100.00	100.12	130.00	130.16
67	4 5 8	104.67	105.55	134.67	135.81
68	4 5 9	128.00	104.48	158.00	128.96
69	4 6 7	104.00	100.51	134.00	129.50
70	4 6 8	108.67	105.75	138.67	134.95
71	4 6 9	132.00	104.67	162.00	128.46
72	4 7 8	113.33	105.83	143.33	133.87
73	4 7 9	136.67	104.79	166.67	127.79
74	4 8 9	141.33	108.97	171.33	132.10
75	5 6 7	102.67	94.62	132.67	122.27
76	5 6 8	107.33	99.58	137.33	127.41
77	5 6 9	130.67	99.64	160.67	122.52
78	5 7 8	112.00	99.91	142.00	126.68
79	5 7 9	135.33	99.91	165.33	122.06
80	5 8 9	140.00	103.91	170.00	126.18
81	6 7 8	116.00	100.27	146.00	126.20
82	6 7 9	139.33	100.21	169.33	121.79
83	6 8 9	144.00	104.10	174.00	125.79
84	7 8 9	148.67	104.23	178.67	125.26
TOTAL		8698.67	8707.67	11218.67	11325.16
EXPECTATION		103.56	103.66	133.56	134.82
VARIANCE		405.32	18.04	405.32	65.11

Verify numerically that the estimator $\hat{y}_{HR}$ is unbiased as an estimator of $\bar{Y}$. (The estimator $\hat{y}_{HR}$ is sometimes called the Hartley-Ross unbiased ratio estimator, after the two statisticians who originally proposed it.)

12.3.7   In a health study, a sample of households was taken in a fairly large city. For each sample household, both the number of persons living there who had had the "flu" during the previous 4 weeks (y_i) and the total number of persons in the household (x_i) were observed. This gave a sample of n pairs of observations (y_i, x_i) $i=1, 2,...,n$.

One objective of the survey was to estimate the overall city sickness rate of Y/X. To do this the estimator y/x, where y and x are sample totals, was proposed. However, the total number of persons in the city X was known from census information and so the estimator $N\bar{y}/X$ was also proposed. The argument was given that the latter estimator should be better because it utilizes the known information on the total, X. Discuss this claim.

12.3.8   Suppose that in the IRS study of taxpayers we wish to estimate the ratio Y/X of actual income to reported income. The sample estimator $\bar{y}/\bar{x} = y/x$ is proposed. Using Table 12.4 calculate $E(\bar{y}/\bar{x})$, $Var(\bar{y}/\bar{x})$ and $MSE(\bar{y}/\bar{x})$. (There is an easy way and a hard way to get these results from Table 12.4.)

The estimator $\bar{y}/\bar{X}$ is a second possible estimator of Y/X. Again using Table 12.4 calculate $E(\bar{y}/\bar{X})$, $Var(\bar{y}/\bar{X})$ and $MSE(\bar{y}/\bar{X})$. (The above remark about ease of calculation also applies here.)

Which estimator do you prefer and why? Compare your answer with the answer you gave to Exercise 12.3.7.

12.3.9   Many manufacturers sell directly to wholesalers and as a result are unaware of the price at which their products are sold to the public. To estimate the average selling price P (total retail revenue/number of units sold) a sample of orders was selected and for each order both the number of items and the total retail revenue were obtained. If the sample sum of the retail revenues was $111,000 and the number of items sold 943 give an appropriate estimator of P. Is it unbiased? What variance formula is appropriate for this estimator?

12.4 Regression Estimation

As we just saw in Section 12.3, the ratio estimator, $\hat{y}_r$, can have an inflated bias and variance if the relationship between y and x does not pass through the origin. This is particularly true if the sample size is small, as it might be within a stratum. Regression estimators, which are discussed in this section, are not sensitive to the failure of the relationship between y and x to pass through the origin. The most popular regression estimator is formed by evaluating the straight line $y = \bar{y} + b(x - \bar{x})$ at the point $\bar{X}$ to give

$$\hat{y}_b = \bar{y} + b(\bar{X} - \bar{x}) \tag{12.8}$$

as an estimator of $\bar{Y}$; see Figure 12.4. In the formula for $\hat{y}_b$, b is the slope of the straight line relating the points y and x and is calculated from the sample by the formula,

$$b = \frac{\sum (y_i - \bar{y})(x_i - \bar{x})}{\sum (x_i - \bar{x})^2} \tag{12.9}$$

As always, $\bar{y}$ and $\bar{x}$ are sample means and $\bar{X}$ is the population mean.

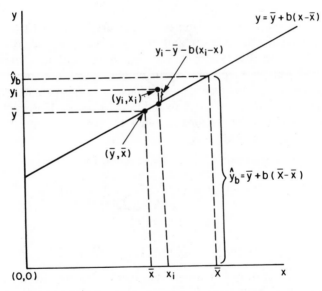

Figure 12.4. The regression estimator, $\hat{y}_b$.

The sampling distribution of $\hat{y}_b$ for samples of three from nine is given in Table 12.4. Notice, that $\hat{y}_b$ is biased by an amount, $101.24 - 103.56 = -2.32$, which is larger than the bias in $\hat{y}_r$ and, that it also has a larger variance than $\hat{y}_r$, 39.90 to 18.04. This makes $\hat{y}_r$ seem better than $\hat{y}_b$. The problem is that the small sample size permits the estimated slope b to vary greatly over the different samples, and this is reflected in both the bias and variance of $\hat{y}_b$. The ratio estimator, $\hat{y}_r$, does not have this problem because the line $y = (\bar{y}/\bar{x})x$ is forced to go through the origin. This makes the slope less variable. This small tax-payer example does *not* show the regression estimator, $\hat{y}_b$, to its best advantage, although it does reveal that $\hat{y}_b$ also has difficulty when the sample sizes are small.

Without abandoning the taxpayer example, one way to illustrate the usual effectiveness of $\hat{y}_b$ is to remove the variability in b. To do this, we use the population value of $B = 1.096$ (calculated by using all nine taxpayers) in place of the sample value, b, which is different for every possible sample. This procedure has about the same effect in removing variability as a large increase in sample size. The substitution of B for b gives a new estimator,

$$\hat{y}_B = \bar{y} + B(\bar{X} - \bar{x}). \qquad (12.10)$$

The sampling distribution of $\hat{y}_B$ appears in Table 12.4. Notice that the bias virtually disappears and that the variance drops to 16.00 in comparison with $Var(\hat{y}_b) = 39.90$. We have said that the bias in $\hat{y}_B$ "virtually" disappears because it can be shown mathematically that $\hat{y}_B$ is usually biased. In this example, the bias in $\hat{y}_B$ is so small that rounding makes it appear that $E(\hat{y}_B)$ is exactly equal to $\bar{Y}$. However, this illus-tration does give a better idea of what to expect when $\hat{y}_b$ is used with reasonably large sample sizes.

The sample variance of $\hat{y}_b$ is calculated by

$$var(\hat{y}_b) = s^2 \left\{ \frac{1}{n} + \frac{(\bar{X} - \bar{x})^2}{\sum(x_i - \bar{x})^2} \right\}, \qquad (12.11)$$

where

$$s^2 = \sum[y_i - \bar{y} - b(x_i - x)]^2/(n - 2), \qquad (12.12)$$

and there are a number of things that can be learned from these formu-las. First, notice that $var(\hat{y}_b)$ gets larger as the term $(\bar{X} - \bar{x})^2$ gets larger; that is, the variance gets quadratically larger the farther $\bar{X}$ is from $\bar{x}$. This is to be expected because the farther the line is evaluated from the location, $\bar{x}$, of the sample, the more variance we expect to have. The second point is that the sample size (3) in our illustration is a *very* small sample on which to compute a regression estimate. In fact, n literally cannot be smaller than 3 because $n - 2$ appears in the divisor of s^2. In practice, the variance of $\hat{y}_b$ is acceptably stable, even with moderately large sample sizes ($N > 20$).

Finally, we remark that it can be shown that the bias in $\hat{y}_b$, similar to that of $\hat{y}_r$, also decreases as the sample size increases. In fact, if n is moderately large, it is actually difficult to create unacceptable biases in $\hat{y}_b$; this is true even if the relationship between y and x is curved rather than a straight line.

In summary, the regression estimator is useful when the straight line describing the relationship between y and x does not pass close to the origin. This is the situation in which the ratio estimator may have both an inflated bias and variance. Then why bother with the ratio estimator at all? Well, first of all, the ratio, $\overline{Y}/\overline{X}$, itself, may be of direct interest. But even when the mean $\overline{Y}$ is being estimated, the ratio estimator is still useful. There are two reasons. The first is that the ratio estimator is simple to calculate and in many applied problems the sample size is large enough so that bias is not a severe problem. The second reason is that the ratio estimator arises naturally in some special situations in multistage sampling. For example, it is useful when unequal size primaries are sampled with equal probabilities (Section 11.5). Unfortunately, this situation is too complicated to discuss in this book.

In this section we have confined ourselves to the discussion of the estimation of $\overline{Y}$. If, instead of $\overline{Y}$, there is interest in estimating the ratio, $\overline{Y}/\overline{X}$, it is often better to use the ratio of the sample means, $\overline{y}/\overline{x}$, rather than the ratio $\overline{y}/\overline{X}$ which involves the known mean $\overline{X}$. As before, this depends upon the magnitude of the correlation between y and x. Again we offer no proof of this statement, but such proof can be found in most textbooks on the mathematics of sampling.

Exercises

12.4.1   Using the same three samples from Table 12.4 that you selected in Exercise 12.3.1, verify the calculations given for $\hat{y}_b$ and $\hat{y}_B$.

12.4.2   A government statistician must estimate the amount of food preservative that has been added to a week's production of 10,000 jars of a food product. The product was made by putting a constant but unknown weight of a "basic" product in each jar. Then unknown but varying amounts of supplementary products were also added to each jar. This supplementary product is known to contain the preservative but the proportions of preservative is not known. The statistician determines the total net weight of the good product by weighing the entire shipment and subtracting the known weight of the empty containers. Then the statistical person draws a random sample of 200 jars and for each jar determines both the weight of the food product and the amount of the nutrient.

i.   Does this situation seem to call for the use of a ratio or regression estimator? Why?

ii.   Show how to use the regression estimator to estimate:
(a) amount of preservative per jar,
(b) the constant amount of basic food product put into the jars before the food supplement was added,
(c) the proportion of the particular food preservative in the supplement.

12.4.3   In Table 12.3, $Var(\hat{y}_r)$ and $Var(\hat{y}_B)$ are approximately the same. Can you explain why?

12.5   Estimation by Post-Stratification

12.5.1   Post-Stratification

In stratified sampling (Chapter 11) the stratum to which a population unit belongs can be identified in advance; this permits a sample of fixed, predetermined size to be drawn from the strata. But in some situations, it is impossible to determine the exact stratum to which a unit belongs until the unit has actually been sampled. This may be true even though population strata *sizes* are known. For example, the religious affiliation of a person cannot usually be ascertained until the individual has been interviewed; but the strata sizes may well be known. More specifically, it may be known that there are M_i units in the *ith* stratum but it may not be known which particular units make up the $\underline{M_i}$ and hence it is impossible to draw a sample of fixed size. Theoretically, it may be possible to sample until a fixed predetermined number in each stratum is selected and then to ignore any excess observations, but this procedure is frequently prohibitively expensive. In such situations a stratified estimator may still be used by placing each sample unit in its appropriate stratum as it is drawn. This is known as *post-stratification*. The post-stratified estimator is discussed in this section.

It is important to understand that in post-stratification the available information about the strata is used in choosing the form of an *estimator* and *not* in the sample selection procedure. In this regard, the simple but usual description "post-stratification" may be somewhat misleading. However, since the strata information used is almost the same in both the post-stratified estimator and stratified sampling, it is not surprising that the results of the two procedures are very much the same. For example, the reduction in sampling variance achieved by the use of the post-stratified estimator is very much influenced by the homogeneity of the units within the strata, just as we discovered for stratified sampling in Chapter 11.

Finally, one last important introductory point. While we have remarked that the post-stratified estimator is clearly of potential use when units cannot be placed into strata before sampling, post-stratified estimators can also be used when, for whatever reason, a stratified sampling was not used. So, if by oversight, we ignore a potentially valuable stratification at the time of the sample selection, the information can still fruitfully be used in the form of the post-stratified estimator.

12.5.2 An Illustration

To illustrate the post-stratified estimator numerically, consider the stratified population of taxpayers discussed in Section 11.6.1. Further, suppose that a random sample of three is selected from the nine taxpayers by equal probability without replacement and that each sample unit is placed in the appropriate stratum *after* selection. As we have seen there are $\binom{9}{3} = 84$ such samples. All 84 samples are listed in Table 12.6 with the sample units placed in the appropriate strata by post-stratification. One of the clearest features of Table 12.6 is that the number of units which fall into each of the two strata varies over the samples. The possibilities are listed in Table 12.7.

TABLE 12.6

SAMPLING DISTRIBUTION OF $\bar{y}_{pst}$

Sample Number	Sampled Taxpayers	$\underline{S}_1$	$\underline{S}_2$	$\bar{y}_1$	$\bar{y}_2$	$\bar{y}_{pst}$
1	1 2 3	1 2 3		66.67		(66.67)
2	1 2 4	1 2 4		75.33		(75.33)
3	1 2 5	1 2 5		74.00		(74.00)
4	1 2 6	1 2	6	66.00	102.00	82.00
5	1 2 7	1 2	7	66.00	116.00	88.22
6	1 2 8	1 2	8	66.00	130.00	94.44
7	1 2 9	1 2	9	66.00	200.00	125.56
8	1 3 4	1 3 4		74.00		(74.00)
9	1 3 5	1 3 5		72.67		(72.67)
10	1 3 6	1 3	6	64.00	102.00	80.89
11	1 3 7	1 3	7	64.00	116.00	87.11
12	1 3 8	1 3	8	64.00	130.00	93.33
13	1 3 9	1 3	9	64.00	200.00	124.44
14	1 4 5	1 4 5		81.33		(81.33)
15	1 4 6	1 4	6	77.00	102.00	88.11
16	1 4 7	1 4	7	77.00	116.00	94.33
17	1 4 8	1 4	8	77.00	130.00	100.56
18	1 4 9	1 4	9	77.00	200.00	131.67
19	1 5 6	1 5	6	75.00	102.00	87.00
20	1 5 7	1 5	7	75.00	116.00	93.22
21	1 5 8	1 5	8	75.00	130.00	99.44
22	1 5 9	1 5	9	75.00	200.00	130.56
23	1 6 7	1	6 7	60.00	109.00	81.78
24	1 6 8	1	6 8	60.00	116.00	84.89
25	1 6 9	1	6 9	60.00	151.00	100.44
26	1 7 8	1	7 8	60.00	123.00	88.00
27	1 7 9	1	7 9	60.00	158.00	103.56
28	1 8 9	1	8 9	60.00	165.00	106.67
29	2 3 4	2 3 4		78.00		(78.00)
30	2 3 5	2 3 5		76.67		(76.67)
31	2 3 6	2 3	6	70.00	102.00	84.22
32	2 3 7	2 3	7	70.00	116.00	90.44
33	2 3 8	2 3	8	70.00	130.00	96.67
34	2 3 9	2 3	9	70.00	200.00	127.78
35	2 4 5	2 4 5		85.33		(85.33)
36	2 4 6	2 4	6	83.00	102.00	91.44
37	2 4 7	2 4	7	83.00	116.00	97.67
38	2 4 8	2 4	8	83.00	130.00	103.89
39	2 4 9	2 4	9	83.00	200.00	135.00
40	2 5 6	2 5	6	81.00	102.00	90.33

Sample Number	Sampled Taxpayers	$\underline{S}_1$	$\underline{S}_2$	$\bar{y}_1$	$\bar{y}_2$	$\bar{y}_{pst}$
41	2 5 7	2 5	7	81.00	116.00	96.56
42	2 5 8	2 5	8	81.00	130.00	102.78
43	2 5 9	2 5	9	81.00	200.00	133.89
44	2 6 7	2	6 7	72.00	109.00	88.44
45	2 6 8	2	6 8	72.00	116.00	91.56
46	2 6 9	2	6 9	72.00	151.00	107.11
47	2 7 8	2	7 8	72.00	123.00	94.67
48	2 7 9	2	7 9	72.00	158.00	110.22
49	2 8 9	2	8 9	72.00	165.00	113.33
50	3 4 5	3 4 5		84.00		(84.00)
51	3 4 6	3 4	6	81.00	102.00	90.33
52	3 4 7	3 4	7	81.00	116.00	96.56
53	3 4 8	3 4	8	81.00	130.00	102.78
54	3 4 9	3 4	9	81.00	200.00	133.89
55	3 5 6	3 5	6	79.002	102.00	89.22
56	3 5 7	3 5	7	79.00	116.00	95.44
57	3 5 8	3 5	8	79.00	130.00	101.67
58	3 5 9	3 5	9	79.00	200.00	132.78
59	3 6 7	3	6 7	68.00	109.33	86.22
60	3 6 8	3	6 8	68.00	116.00	89.33
61	3 6 9	3	6 9	68.00	151.00	104.89
62	3 7 8	3	7 8	68.00	123.00	92.44
63	3 7 9	3	7 9	68.00	158.00	108.00
64	3 8 9	3	8 9	68.00	165.00	111.11
65	4 5 6	4 5	6	92.00	102.00	96.44
66	4 5 7	4 5	7	92.00	116.00	102.67
67	4 5 8	4 5	8	92.00	130.00	108.89
68	4 5 9	4 5	9	92.00	200.00	140.00
69	4 6 7	4	6 7	94.00	109.00	100.67
70	4 6 8	4	6 8	94.00	116.00	103.78
71	4 6 9	4	6 9	94.00	151.00	119.33
72	4 7 8	4	7 8	94.00	123.00	106.89
73	4 7 9	4	7 9	94.00	158.00	122.44
74	4 8 9	4	8 9	94.00	165.00	125.56
75	5 6 7	5	6 7	90.00	109.00	98.44
76	5 6 8	5	6 8	90.00	116.00	101.56
77	5 6 9	5	6 9	90.00	151.00	117.11
78	5 7 8	5	7 8	90.00	123.00	104.67
79	5 7 9	5	7 9	90.00	158.00	120.22
80	5 8 9	5	8 9	90.00	165.00	123.33
81	6 7 8		6 7 8		116.00	(116.00)
82	6 7 9		6 7 9		139.33	(139.33)
83	6 8 9		6 8 9		144.00	(144.00)
84	7 8 9		7 8 9		148.67	(148.67)
EXPECTATION				76.80	137.00	103.56 (101.96)

TABLE 12.7

DISTRIBUTION OF SAMPLES IN POST-STRATIFICATION

Strata Sample Size		Number of Such Samples
$\underline{S}_1$	$\underline{S}_2$	
0	3	4
1	2	30
2	1	40
3	0	10

Initially, we shall consider the 70 samples in which *both* strata are represented in the sample. For these samples, the post-stratified estimator is formulated as,

$$\bar{y}_{pst} = \sum_{i=1}^{N} \frac{M_i}{M} \bar{y}_i \qquad (12.13)$$

which is exactly the same formula used for stratification in Chapter 11. So for example, in sample 4 in Table 12.6

$$\bar{y}_{pst} = \frac{5}{9}(66.00) + \frac{4}{9}(102.00) = 82.00.$$

This calculation is given for the 70 relevant samples in Table 12.6.

Notice first that, averaged over the 70 samples in which *both* strata are represented, (and *not* over all 84 samples)

$$E(\bar{y}_{pst}) = \frac{1}{70}[82.00 + 88.22 + \cdots + 123.33] = 103.56 \ (12.14)$$

This means that $\bar{y}_{pst}$ is unbiased if we consider *only* samples in which *both* strata are actually represented in the sample. This is sometimes called conditional unbiasedness.

Next, what is the sampling variance of $\bar{y}_{pst}$ over the 70 samples? Recall (Chapter 11) that when $m_1 = 2$ and $m_2 = 1$ that $Var(\bar{y}_{st})$ is equal to 300.43 and that when $m_1 = 1$, $m_2 = 2$ (which is optimum allocation) $Var(\bar{y}_{st})$ is equal to 146.21. Since $\bar{y}_{pst}$ includes both cases in which $m_1 = 2$, $m_2 = 1$ and cases in which $m_1 = 1$, $m_2 = 2$, we might expect that $Var(\bar{y}_{pst})$ would turn out to be somewhere between 146.21 and 300.43. This is indeed the case. Direct calculation of the sampling variance reveals that

$$Var(\bar{y}_{pst}) = \frac{1}{70}[(82.00 - 103.56)^2 + (88.22 - 103.56)^2 + \cdots + (123.33 - 103.56)^2]$$

$$= 233.41. \tag{12.15}$$

So far we have considered only samples in which both strata are represented. Next consider *all* of the samples, including those in which either one of the strata is *not* represented in the sample. From Table 12.7 we see that $\underline{S}_1$ is not represented in 4 samples and $\underline{S}_2$ is not represented in 10. Now we already know from the study of the 70 samples that if the mean of a population which includes only those strata which *are* represented in the sample is to be estimated, then $\bar{y}_{pst}$ is unbiased. This means that if some strata are not represented in the sample then the post-stratified estimator is an unbiased estimator of the mean of the restricted population which does not include the missing strata. But what if we wish to estimate the mean of the whole population including those strata which have not contributed units to the sample? That is, is $\bar{y}_{pst}$ unbiased over all 84 samples? Intuitively, one would suspect that $\bar{y}_{pst}$ is not unbiased in this case and indeed it is not.

To see that $\bar{y}_{pst}$ is biased for $\bar{Y}$ when samples are included which omit some strata, we must first consider the definition of $\bar{y}_{pst}$. The problem is that for those strata which have contributed no sample units there is no sample estimate $\bar{y}_i$ of $\bar{Y}_i$ for use in the formula of $\bar{y}_{pst}$. For these strata there is no statistical way of estimating the stratum mean. However, in the absence of other information, a simple and commonlyused approach is to calculate $\bar{y}_{pst}$ for the represented strata and use this as an estimator of $\bar{Y}$. In our taxpayer example, this procedure implies that the mean of the one represented stratum is used as $\bar{y}_{pst}$. For example in sample number 1, $\bar{y}_{pst} = 66.67$, the sample mean in $\underline{S}_1$. The values of $\bar{y}_{pst}$ for the 14 samples with "missing" strata are given in Table 12.6 from which it can be easily verified that

$$E(\bar{y}_{pst}) = \frac{1}{84}(66.67 + 75.33 + \cdots + 148.67) = 101.96. \tag{12.16}$$

This calculation confirms our initial suspicion that $\bar{y}_{pst}$ is biased over all 84 samples.

Finally what is the variance of $\bar{y}_{pst}$ over all 84 samples? By direct calculation,

$$Var(\bar{y}_{pst}) = \frac{1}{84}[(66.67-103.56)^2 + (75.33-103.56)^2 + \cdots + (148.67-103.56)^2]$$

$$= 342.73 \qquad\qquad (12.17)$$

and so,

$$MSE(\bar{y}_{pst}) = 342.73 + (-1.6)^2 = 345.29 \qquad (12.18)$$

An immediate question is: Why is the variance of $\bar{y}_{pst}$ higher over 84 samples that it is over 70? The answer is that the 14 additional samples are samples in which all units come from a single stratum. And, since S_1 contains the smaller taxpayers and S_2 the larger ones, these samples tend to have means that are farther from the overall population mean than the 70 samples. Thus a sample with units from only one stratum is pushed to one extreme or the other. However the bias effects of a missing stratum depend entirely on the characteristics of that stratum and so in general, it is not possible to reach conclusions about the bias unless something is known about the mean of that missing stratum.

12.5.3 General Remarks

What general statements can be made about post-stratification? First, if only samples are considered in which all strata are represented, $\bar{y}_{pst}$ is unbiased. Furthermore, $\bar{y}_{pst}$ will have a sampling variance which is reduced (in comparison with unrestricted random sampling) in proportion to the differences among stratum means. This is just as in stratified sampling. Furthermore, $Var(\bar{y}_{pst})$ will be larger than $Var(\bar{y}_{st})$ when the stratified sampling uses optimum allocation of the units to strata. However, $Var(\bar{y}_{pst})$ is not as large as $Var(\bar{y}_{st})$ in which a poor allocation of units to strata is used. In fact as we have seen, $Var(\bar{y}_{pst})$ is an average (almost) of $Var(\bar{y}_{st})$ over the various possible allocations of sample units to strata. For example, we saw in the taxpayer example that $Var(\bar{y}_{pst})$ was a value (233.41) between the values of $Var(\bar{y}_{st})$ when $m_1 = 1$, $m_2 = 2$ (300.43), and when $m_2 = 2$, $m_1 = 1$ (146.21). It turns out that it is convenient and reasonably accurate to think of $Var(\bar{y}_{pst})$ as approximately equal to $Var(\bar{y}_{st})$ when *proportional* allocation is used. The intuitive logic is that the sample can be excepted to spread itself over the strata in proportion to the strata sizes.

Finally, what general statements can be made about post-stratification if some strata are not represented and if one wishes to make inferences about the mean of the entire population including those which are not represented? Unfortunately in this case, as we have seen, $\bar{y}_{pst}$ is biased and the bias depends directly upon how different the means of the missing strata are from the rest of the strata. If nothing is known about this difference, nothing is known about the bias. It could be arbitrarily large.

As an approximate estimate of $Var(\bar{y}_{pst})$ use the formula for $var(\bar{y}_{st})$ described in Section 11.6.1.

Exercises

12.5.1   Show by combinatorial argument that the numbers of samples of the four types are as given (in Table 12.7). Also, verify the result by direct counting in Table 12.6.

12.5.2   Given that
$$(82.00)^2 + (88.22)^2 + \cdots + (123.33)^2 = 767,066.18, \quad \text{verify}$$
that $Var(\bar{y}_{pst}) = 233.41$ (Equation 12.15) as given in Section 12.5.2.

12.5.3   In the taxpayer example of Table 12.6 confirm that $E(\bar{y}_{pst}) = 101.96$ over all 84 samples, and hence that $\bar{y}_{pst}$ has a bias of -1.6.

12.6 Technical Biases

We have just seen that in general both the ratio estimator, $\hat{y}_r$, and the regression estimator, $\hat{r}_b$, are biased. These biases occur because the algebraic form of the estimator does not result in an expectation which is equal to the true value of $\bar{Y}$. Such a bias is called a technical bias. It is important to keep in mind that technical biases are different from the biases caused by measurement errors (Chapter 6) and the biases caused by distortion of the selection probabilities (Chapter 10).

In many cases, the technical biases of $\hat{r}_r$ and $\hat{r}_b$ are unimportant. These estimators have been studied mathematically and for both the bias gets smaller as the sample size gets larger. Generally speaking, $\hat{y}_r$ and $\hat{r}_b$ are useful estimators, the main situation in which care must be taken is with small sample sizes within strata.

Exercises

12.6.1   Calculate the bias and mean square error of the estimators $\bar{y}$, $\hat{r}_r$, $\hat{r}_b$, and $\hat{r}_B$ in Table 12.4. Discuss the comparison of these mean square errors.

12.6.2   Calculate the bias and mean square error of the estimators $\bar{y}$ and $\hat{r}_r$ in Table 12.5 (two cases for each estimator). Discuss the comparison of these mean square errors.

12.6.3   Compare the mean square error of the estimator $r\bar{X}$ (Exercise 12.3.5) with the mean square errors calculated in Exercise 12.6.1. Discuss these comparisons.

IMPORTANT NEW IDEAS

robust estimators   auxiliary information
correlation   ratio estimator
regression estimator   technical bias
post-stratified estimator

CHAPTER 13

THE VAGUE ANSWER TO THE PRECISE QUESTION OF SAMPLE SIZE

13.1 Diminishing Returns from Bigger Samples

Bigger sample sizes make the sampling distribution of the mean bunch up, but not proportionally. To illustrate, suppose that more staff is made available for auditing taxpayers, so that the sample size can be increased from 2 of 8 to 3, 4, and even 5 of 8. If $n = 2$, and sampling is without replacement, we saw in Chapter 5 that there are $\binom{8}{2} = 28$ possible samples. Similarly, if $n = 3$, there are $\binom{8}{3} = 56$ possible samples, and if $n = 4$ there are $\binom{8}{4} = 70$, and if $n = 5$ there are $\binom{8}{5} = 56$. Histograms of the four sampling distributions for the sample sizes $n = 2, 3, 4,$ and 5, are shown in Figure 13.1. Since the horizontal scales are the same for each histogram, it is clear that the sampling distributions are bunching up in the middle with increasing n. This bunching is a result of the sampling variances decreasing with increasing n; the calculated variances are shown in Figure 13.1.

If sampling is without replacement from the taxpayer population,

$$Var\,(\bar{y}) = (1 - \frac{n}{8})\,\frac{589.4}{n} = \frac{589.4}{n} - 73.7,$$

for a general value of n ($1 \leqslant n \leqslant 8$). This variance is plotted as a function of n in Figure 13.2. Similarly, if sampling is with replacement,

$$Var\,(\bar{y}) = \frac{515.8}{n},$$

which is also plotted as a function of n in Figure 13.2.

From Figure 13.2, it is clear that increasing the sample size has the desirable effect of reducing sampling variance. But, it is also clear that the reduction diminishes progressively. When sampling with

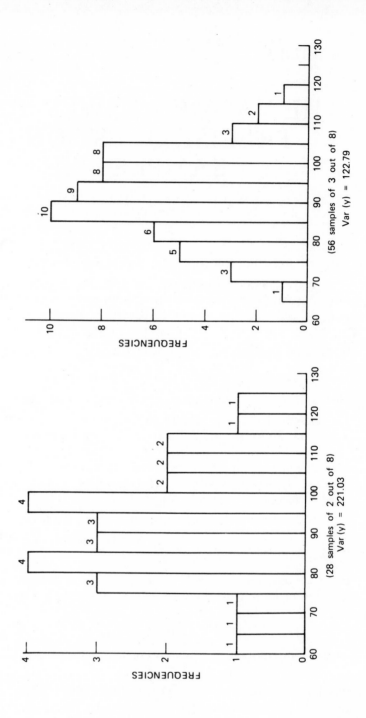

212

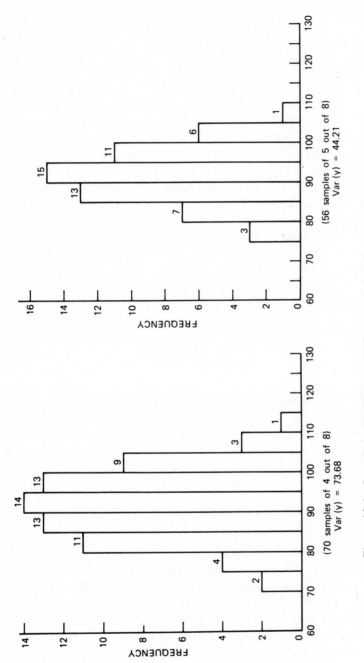

Figure 13.1. Sampling distribution of $\bar{y}$ for $n = 2, 3, 4, 5$ taxpayer population.

213

replacement, increasing the sample size from 1 to 2 cuts the variance in half, from 515.8 to $(515.8)/2 = 257.9$. *But,* the next halving of the variance, requires an increase in sample size of not one but *two,* to $n = 4$. In general, to cut the variance in half, you must double the sample size. This is one of the reasons why the biggest gains in efficiency are often made by clever sampling designs and estimators, rather than by increased sample size.

Furthermore, while the variance decreases in proportion to $1/n$ the reduction in the width of confidence intervals is even slower. Recall from Chapters 7 and 8 that the width of confidence intervals depends upon the square root of the variance rather than the variance itself. Hence the confidence intervals decrease in proportion to $1/\sqrt{n}$ and not $1/n$. Finally, keep in mind that if sampling is without replacement and $n = N$, then $Var(\bar{y}) = 0$ and the confidence intervals will collapse to a single point.

Exercises

13.1.1   Verify the calculation of the four variances given in Figure 13.1.

13.2 A Mathematically Precise Sample Size

13.2.1 Trial and Error

Suppose that the IRS analysts decide that to be practically useful, the estimated taxpayer mean should be within 10 (thousand dollars) of the true mean, $\bar{Y} = 91.50$, that is, within the range 81.5 to 101.5. What are the implications of such a requirement?

First consider the sampling distribution for $n = 4$, Table 13.1. Seven of the 70 possible sample means lie below 81.5, and seven lie above 101.5. This can be written formally as,

$$Prob\{-10 < \bar{y} - 91.50 < +10\} = 1 - \frac{14}{70} = 0.80.$$

Next by manipulating the inequalities in the preceding expression in *exactly* the same way as in Chapters 7 and 8, the expression immediately above can be written as

$$Prob\{\bar{y}-10 < 91.50 < \bar{y}+10\} = 0.80.$$

This is, of course, a statement of 80% confidence intervals. Specifically, the confidence interval calculated by $\{\bar{y}-10, \bar{y}+10\}$ will cover the true mean, $\bar{Y} = 91.50$, exactly 80% of the time. Thus, with a sample size of $n = 4$, there is a 20% risk that the sample mean will *not* be within the interval 81.5 to 101.5. If this risk is to be lowered (without changing to a different sampling design and/or estimator), the sample size will have to be increased. This will cause the sampling distribution to bunch up in the middle and more of it will be included in the interval 81.5 to 101.5. To see how this works, suppose that the IRS analysts feel that the 20% risk is too large and that it should be reduced to 12.5%. How much bigger must the sample size be to achieve this level of risk?

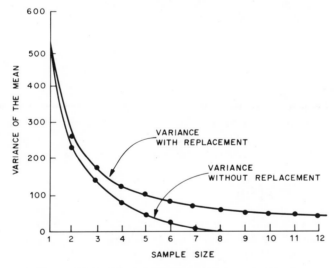

Figure 13.2. Diminishing returns from increasing
sample size (taxpayer population $n = 8$).

With a sample size of $n = 4$, the sample mean will be outside of the specified accuracy interval, $\{81.5, 101.5\}$, 20% of the time. This risk of 20% is too large and is to be lowered to 12.5%. To do this requires a sampling distribution which is more bunched up in the middle. Therefore, consider the sampling distribution based on samples of five observations, Table 13.2. Four of the possible means lie below 81.5 and three are above 101.5. So 7 of the possible 56 means lie outside of the specified limits of {81.5, 101.5}. This can be stated as,

$$Prob\{-10 < \bar{y} - 91.50 < +10\} = 1 - \frac{7}{56} = \frac{49}{56} = 0.875$$

or equivalently,

$$Prob\{\bar{y} - 10 < 91.50 < \bar{y} + 10\} = 0.875.$$

This means that the intervals {$\bar{y} - 10$, $\bar{y} + 10$} cover the true mean, 91.50, in exactly the specified fraction (0.875) of cases. Consequently, we see that a sample of size 5 is big enough to meet the more stringent IRS requirements of sample accuracy.

13.2.2 A Formula for Sample Size

In the example of Section 13.2.1, the determination of 5 as a satisfactory sample size was achieved by trial and error. Two different sample sizes were tried and it was found that $n = 5$ was sufficient. Clearly, this approach would be cumbersome in practice. Consequently, it would be very useful to have a formula for selecting a sample size. Such a formula is found in this section.

In Section 13.2.1, it was specified that the IRS analysts wanted to estimate the true mean ($\bar{Y} = 91.5$) to within ± 10. We also saw in that section that there was some risk that the actual sample mean would not be within the desired limits. An acceptable risk of this actually happening was taken to be 0.125, or 1 chance in 8, which is succinctly written as,

$$Prob\{-10 < \bar{y} - 91.50 < +10\} = 0.875. \tag{13.1}$$

Equation 13.1 is very similar to expressions that appeared often in Chapters 7 and 8 on confidence intervals and the normal distribution. In fact, the central limit approximation says that,

$$Prob\{-1.53\sqrt{Var(\bar{y})} < \bar{y} - 91.50 < 1.53\sqrt{Var(\bar{y})}\} = 0.875, \tag{13.2}$$

where ± 1.53 are the normal quantiles which cut off 12.5% of the area (6.25% at each tail) in the tails of the normal distribution.

TABLE 13.1

SAMPLING DISTRIBUTION OF $\bar{y}$

[4 taxpayers from 8]

Sample Number	Sampled Taxpayers	Sample $\bar{y}$	Sample Number	Sampled Taxpayers	Sample $\bar{y}$
1	1 2 3 4	73.50	38	2 3 4 7	87.50
2	1 2 3 5	72.50	39	2 3 4 8	91.00
3	1 2 3 6	75.50	40	2 3 5 6	83.00
4	1 2 3 7	79.00	41	2 3 5 7	86.50
5	1 2 3 8	82.00	42	2 3 5 8	90.00
6	1 2 4 5	79.00	43	2 3 6 7	89.50
7	1 2 4 6	82.00	44	2 3 6 8	93.00
8	1 2 4 7	85.50	45	2 3 7 8	96.50
9	1 2 4 8	89.00	46	2 4 5 6	89.50
10	1 2 5 6	81.00	47	2 4 5 7	93.00
11	1 2 5 7	84.50	48	2 4 5 8	96.50
12	1 2 5 8	88.00	49	2 4 6 7	96.00
13	1 2 6 7	87.50	50	2 4 6 8	99.50
14	1 2 6 8	91.00	51	2 4 7 8	103.00
15	1 2 7 8	94.50	52	2 5 6 7	95.00
16	1 3 4 5	78.00	53	2 5 6 8	98.50
17	1 3 4 6	81.00	54	2 5 7 8	102.00
18	1 3 4 7	84.50	55	2 6 7 8	105.00
19	1 3 4 8	88.00	56	3 4 5 6	88.50
20	1 3 5 6	80.00	57	3 4 5 7	92.00
21	1 3 5 7	83.50	58	3 4 5 8	95.50
22	1 3 5 8	87.00	59	3 4 6 7	95.00
23	1 3 6 7	86.50	60	3 4 6 8	98.50
24	1 3 6 8	90.00	61	3 4 7 8	102.00
25	1 3 7 8	93.50	62	3 5 6 7	94.00
26	1 4 5 6	86.50	63	3 5 6 8	97.50
27	1 4 5 7	90.00	64	3 5 7 8	101.00
28	1 4 5 8	93.50	65	3 6 7 8	104.00
29	1 4 6 7	93.00	66	4 5 6 7	100.50
30	1 4 6 8	96.50	67	4 5 6 8	104.00
31	1 4 7 8	100.00	68	4 5 7 8	107.50
32	1 5 6 7	92.00	69	4 6 7 8	110.50
33	1 5 6 8	95.50	70	5 6 7 8	109.50
34	1 5 7 8	99.00			
35	1 6 7 8	102.00	TOTAL		6405.00
36	2 3 4 5	81.00	MEAN		91.50
37	2 3 4 6	84.00	VARIANCE		73.69

TABLE 13.2

SAMPLING DISTRIBUTION OF $\bar{y}$

[5 taxpayers from 8]

Sample Number	Sampled Taxpayers	Sample $\bar{y}$	Sample Number	Sampled Taxpayers	Sample $\bar{y}$
1	1 2 3 4 5	76.80	29	1 3 5 7 8	92.80
2	1 2 3 4 6	79.20	30	1 3 6 7 8	95.20
3	1 2 3 4 7	82.00	31	1 4 5 6 7	92.40
4	1 2 3 4 8	84.80	32	1 4 5 6 8	95.20
5	1 2 3 5 6	78.40	33	1 4 5 7 8	98.00
6	1 2 3 5 7	81.20	34	1 4 6 7 8	100.40
7	1 2 3 5 8	84.00	35	1 5 6 7 8	99.60
8	1 2 3 6 7	83.60	36	2 3 4 5 6	85.20
9	1 2 3 6 8	86.40	37	2 3 4 5 7	88.00
10	1 2 3 7 8	89.20	38	2 3 4 5 8	90.80
11	1 2 4 5 6	83.60	39	2 3 4 6 7	90.40
12	1 2 4 5 7	86.40	40	2 3 4 6 8	93.20
13	1 2 4 5 8	89.20	41	2 3 4 7 8	96.00
14	1 2 4 6 7	88.80	42	2 3 5 6 7	89.60
15	1 2 4 6 8	91.60	43	2 3 5 6 8	92.40
16	1 2 4 7 8	94.40	44	2 3 5 7 8	95.20
17	1 2 5 6 7	88.00	45	2 3 6 7 8	97.60
18	1 2 5 6 8	90.80	46	2 4 5 6 7	94.80
19	1 2 5 7 8	93.60	47	2 4 5 6 8	97.60
20	1 2 6 7 8	96.00	48	2 4 5 7 8	100.40
21	1 3 4 5 6	82.80	49	2 4 6 7 8	102.80
22	1 3 4 5 7	85.60	50	2 5 6 7 8	102.00
23	1 3 4 5 8	88.40	51	3 4 5 6 7	94.00
24	1 3 4 6 7	88.00	52	3 4 5 6 8	96.80
25	1 3 4 6 8	90.80	53	3 4 5 7 8	99.00
26	1 3 4 7 8	93.60	54	3 5 6 7 8	102.00
27	1 3 5 6 7	87.20	55	3 5 6 7 8	101.20
28	1 3 5 6 8	90.00	56	4 5 6 7 8	106.40

TOTAL	5124.00
MEAN	91.50
VARIANCE	44.21

Consequently, if we *want* Equation 13.1 to be true and are willing to assume that Equation 13.2 *is* true, we can arrange it by setting

$$1.53\sqrt{Var(\bar{y})} = 10. \tag{13.3}$$

This equality ensures that Equations 13.1 and 13.2 are the same. Furthermore, by squaring and setting $Var(\bar{y}) = \sigma^2/n$, which is the variance formula for sampling with replacement, or without replacement with small n/N, then Equation 13.3 can be written as,

$$(1.53)^2 \frac{\sigma^2}{n} = (10)^2,$$

or upon solving for n,

$$n = (\frac{1.53}{10})^2 \sigma^2. \tag{13.4}$$

Here is a formula for n. In this case, the value of n obtained from the formula indicates that the sample mean will be within the interval 81.5 to 101.5, 87.5% of the time.

In general notation, if the desired accuracy of the sample mean is ·denoted by d and the risk by α, then the formula for n is

$$n = (\frac{Z_{\alpha/2}}{d})^2 \sigma^2 \tag{13.5}$$

where $Z_{\alpha/2}$ is the normal quantile which cuts off $\alpha/2$ of the area of the normal distribution in the upper tail.

In the taxpayer example, $\sigma^2 = 515.8$ which gives,

$$n = \frac{(1.53)^2(515.8)}{100} = 12.1.$$

This indicated sample size is larger than the size of the whole taxpayer population, $(N = 8)$. In practice, with larger samples and populations, the formula works reasonably well. But as we pointed out in Section 10.1, when sampling with replacement, the possibility of duplicates does mean that sample sizes may theoretically be larger than the population size N.

When sampling without replacement and the sampling fraction n/N is not negligible, the formula for sample size can be improved by including the finite population correction in the formula for $Var(\bar{y})$ in Equation 13.3. So instead of substituting $Var(\bar{y}) = \sigma^2/n$ as we did previously, substitute $Var(\bar{y}) = (1 - \frac{n}{N}) \frac{S^2}{n}$ to obtain

$$(1.53)^2(1 - \frac{n}{N}) \frac{S^2}{n} = 10^2 \tag{13.6}$$

or in general notation,

$$Z_{\alpha/2}^2 \left(1 - \frac{n}{N}\right) \frac{S^2}{n} = d^2. \tag{13.7}$$

In the taxpayer example, Equation 13.6 is equal to

$$(1.53)^2 \left(\frac{1}{n} - \frac{1}{8}\right) 589.4 = 100, \tag{13.8}$$

in which the only unknown value is n, the sample size. The reader can verify that $n = 5.063$ is the solution of Equation 13.8. The closest integer sample size is $n = 5$. Now this is a very interesting result because in the previous subsection (13.2.1) the value of $n = 5$ was found to be *exactly* the correct value of n. So, in this case, the formula has given an unusually accurate result, especially considering that the involved approximations should not be particularly good for such small values of n and N.

The reader can easily verify that, if n_0 is the sample size determined by Equation 13.4 or 13.5, which does not include the finite population correction, then

$$n = \frac{n_0}{1 + n_0/N} \tag{13.9}$$

is the formula when the finite population correction is included.

From the preceding derivation of formulas for sample size, one can easily get the impression that finding a required sample size is a precise operation which involves merely solving a simple equation. Unfortunately, as we shall see in the next section, it isn't quite that easy; major difficulties do exist.

Before finishing this section, it is worthwhile making a summary of our findings. We have seen that estimation of sample size requires that three things be specified. All three of these were used in the examples just discussed. They are:

(1)   a required precision, d, that expresses how close to the true mean the sample mean should be;

(2)   a measure of the variability in the population, (σ^2 and S^2 were both used in the examples);

(3)   a specification of the acceptable risk that the actual confidence interval does *not* cover the true mean. In the example, this risk was equal to 0.125.

These three requirements are discussed below in order.

Clearly, there must be some statement of precision required in the study. If there is no such statement, any sample size will do. In the taxpayer example, mean income was to be known to within ±10. In addition, something needs to be known about the dispersion of the target population. Dispersion influences the sample size requirements. In the extreme, if everyone in a certain golf club is known to have the same age a sample of 1 would estimate their average age. The dispersion in this population is zero and, consequently, one observation tells all. While populations with zero dispersion are not very interesting in practice, the extremity of the example makes it clear that the sample accuracy does depend upon population dispersion.

Finally, every sample is subject to the vagaries of random sampling, and no matter what precision is specified, the actual sample mean may turn out to be a long way from the true mean. Consequently, to determine sample size, a statement of this risk must also be made. In the taxpayer example, this risk was 0.125. This means that in one sample of eight, the sample mean will be farther away from the true mean than the specified accuracy.

A misleading feature of our small example is that if $n = N = 8$, then $\bar{y} = \bar{Y} = 91.5$. In this case there is no risk, the sample mean is exactly equal to the true mean. So it is well to keep in mind that in real studies, we shall be dealing with samples and not censuses, and that n will usually be much smaller than N. In practice, the risk that the sample mean is farther away from the true mean than the specified amount can never be reduced to zero.

Exercises

13.2.1 Verify numerically (use Table 13.1) that the statement given in Section 13.2 that

$$Prob\{-10 < \bar{y} - 91.5 < +10\} = 0.80$$

is correct.

13.2.2 Verify numerically (use Table 13.1) that the statement given in Section 13.2 that

$$Prob\{\bar{y} - 10 < 91.5 < \bar{y} + 10\} = 0.80$$

is correct.

13.2.3   Show algebraically that the two equations given in the Exercises 13.2.1 and 13.2.2 are equivalent.

13.2.4   Verify either numerically or algebraically that the following two equations (and which appear in Section 13.2) are equivalent.

$$Prob\{-10 < \bar{y} - 91.5 < 10\} = 0.875$$

$$Prob\{\bar{y} - 10 < 91.5 < \bar{y} + 10\} = 0.875.$$

13.2.5   If a sample is drawn from a normal distribution with $\sigma = 5$, find by trial and error, the sample size required to estimate the true mean to within ± 2.

13.2.6   If a sample is drawn from a normal population with $\sigma = 5$, find by trial and error, the sample size required to estimate the true mean to within $\pm 20\%$ of the true mean. Do two cases. In one case assume that $\bar{Y}$ is approximately 5 and in the other assume that $\bar{Y}$ is approximately 25. Compare and discuss the two results.

13.2.7   Solve the problems described in Exercises 13.2.5 and 13.2.6 by use of the formula of Equation 13.5 rather than by trial and error.

13.2.8   Verify algebraically that the formula for sample size is (Equation 13.9)

$$n = \frac{n_0}{1 + \dfrac{n_0}{N}},$$

where n_0 is the sample size indicated without inclusion of the finite population correction.

13.2.9   The population of the heights of all United States adult males has a standard deviation of about 3 inches. What size sample should be taken to give a 90% confidence interval of length 1 inch for $\bar{Y}$, the average height of United States males? What size samples are necessary for a 99% confidence interval of length of 1 inch and a 99% confidence interval of length 1/2 inch?

13.2.10   As discussed in Chapter 13 there are a number of approximations and even guesses involved in the estimation of sample size. Sometimes in practice, it is assumed that the sample proportion p is approximately normally distributed with variance

PQ/n. Derive a sample size n *adequate* to estimate the true proportion P to an accuracy of $\pm 3\%$ with a confidence of 95%. Ignore finite population corrections.

13.2.11 A random sample of cereal boxes was drawn to determine if the cereal weight differed significantly from the stated 12 oz. The sample mean was 12.229 oz and the sample standard deviation was 0.155 oz. How large should the sample size have been to estimate the mean weight to within 15/100 of an ounce with 95% confidence?

   The percentage of boxes with cereal weight less than 12 oz was 5.25. How large a sample is required to estimate the percentage of "underweight" boxes to within 1% with a 95% confidence?

13.2.12 Analysts want to know the proportion of people in the United States who suffer from hypertension. This proportion is believed to be less than 0.15. How large a sample should be drawn to estimate the true proportion to within 0.02 with at most a 0.05 chance of being in error?

13.3 The Guesswork

In the framework of Sections 13.1 and 13.2, selection of a sample size appears to be a problem of formal mathematics, that is, solving a formula. In this section, we are going to dig a little deeper and find that, unfortunately in actual practice, there is both approximation and guesswork involved in sample size selection.

One approximation in the formulas for sample size has already been mentioned. It is the central limit theorem. To develop Equation 13.5 it was assumed that $\bar{y}$ is approximately normally distributed. This approximation was discussed in Chapter 8, and fortunately, in many applications, works reasonably well. Nevertheless, this is approximation number one.

Next, an obvious difficulty is that σ^2, the variance of the population is required in the formulas for n. Unfortunately, this variance is rarely known in practice. Even *after* the sample is drawn only an estimate of σ^2 is available, and this is too late to help in sample size estimation which must be done *before* the study is conducted.[1] One

---

1. Sequential sampling is sometimes useful for this problem, but this technique is not covered in this book.

practical solution is to obtain an estimate of variance from an earlier study, or perhaps even from a pilot study. Sometimes, such information is quite good enough to use in the formula for n as a substitute for the true value of σ^2.

If no variance estimate is available from a pilot or earlier study, there is a range approximation that may be useful. Determine the largest and smallest observations that have been seen, recorded, or imagined. This gives a crude estimate, $\hat{R}$, of the range, R. In normal samples of about 200 to 1000 in size, the average value of the sample range is 6σ. So, a crude estimate of σ can be obtained by dividing $\hat{R}$ by 6. This range technique may be better than no estimate at all, but in any event, the substitution of an estimate or guess of the population variance is approximation number two.

A third difficulty in the process of picking a sample size is that almost all surveys are multivariate. Usually many measurements, for example height, weight, income, age, are made on each sampled unit, and each of these will result in a *different* specification of sample size. How are these different sample sizes to be reconciled? Unfortunately, there are no good methods. There are statistical formulas for combining different estimates into a single sample size, but these ultimately involve an arbitrary weighting together of the various measurements. In practice, a reasonable approach is to estimate a sample size for each of the measurements separately, and then try to reconcile the largest sample requirements with the budget and accuracy constraints. With luck, some acceptable arbitration can take place. However, such arbitration does add a completely subjective aspect to the process of picking a sample size. So, add this approximation to the two earlier ones, for a total of three.

Exercises

13.3.1   A survey was conducted of various industrial groups to study the differences in sizes of firms in the groups. For simplicity, suppose that there are four industrial groups with the following numbers of firms in each group: $I_1 = 8000$, $I_2 = 5000$, $I_3 = 3000$, $I_4 = 1000$.

(a) If there is no prior variance information available and there is enough money to interview 250 firms, how many firms should be interviewed from each class if we wish to estimate the average number of employees for all 17,000 firms?

(b) Given the same conditions as part (a), how many firms should be interviewed from each class if we wish to estimate the mean number of employees in each industrial group with equal accuracy.

13.3.2   A bus company wishes to estimate the average passenger waiting time at one of its stops. The proposal is to go to the stop at randomly selected times of the day and observe the time until the arrival of the next bus. Discuss this sampling scheme including possible estimators.

13.3.3   A lumber company wishes to obtain an estimate of the daily volume of lumber handled by each of several saw mills. The proposal is to sample the log which is just passing a specified point on a conveyor belt. This would be repeated every 15 minutes. Each sample log would be removed from the belt and measured. The total volume of lumber could then be estimated by multiplying the mean log volume of the sample by the total number of logs. Discuss this scheme.

13.4 Sample Size, the Last Question Not the First

Invariably, sample size is the first question asked in the design of a new survey. It is an important question, and one that must be answered. However, the problem of choosing the number of observations is in some ways one of the most difficult in applied statistics. It takes technical knowledge and experience to approach the problem properly. Unfortunately to laymen, the problem of selecting a sample size may appear to be an easy one; one that statisticians deal with routinely many times each day. The problem is not routine. In fact, its deceptively simple appearance has probably driven more than one statistician to fill a hat with a selection of possible n values, and hide it in an inconspicuous, but accessible, place in his office. A snappy "n=381" can leave a client very impressed, even if badly informed. To further encourage this, an answer of 381 may never be subsequently found wanting because the problems of analysis will eventually occupy almost all of the client's best thinking.

Sample size estimation is one of the *last* questions to be answered, not the first. Why? Recall that in the taxpayer illustration, a sample of size 5 was found to be adequate. But notice that only the sample mean,

$\bar{y}$, was considered as an estimator. Similarly, only simple random sampling was considered. If a different estimator is used (say a ratio estimator), and/or a different sampling design (say a cluster sample), the sampling variance formula would be different and a different sample size would achieve the desired accuracy. The formal estimation of sample size *requires* specification of both an estimator *and* the method of sampling. These must be considered first.

Why then does the sample size issue tend to arise so early? Clearly budget considerations play a role. The client wants to know early what can be accomplished within his available time and budget. However, looked at this way, the problem changes, it becomes one of sample *allocation* and not simply one of sample size estimation. The problem is to use the budget in the most efficient way, which clearly means choosing both good estimators and good methods of sample selection.

Unfortunately, even if all the preceding problems disappeared, the actual sample accuracy will be different from the anticipated sample accuracy. The reason is that some bias invariably affects the sample. This bias may be the result of technical factors, data errors, or selection errors -- most likely the latter two, but in any event, such biases ensure that the final accuracy is different from the specified accuracy. It is important to realize that sample size specification merely gives guidelines for assisting intelligent decisions in the conduct of the survey.

Finally, there is one last important point about sample size and finite population corrections. The complaint is often popularly heard that small samples of 1000 to 2000 people are too small to be used to estimate characteristics of the entire United States population of over 200 million people. Actually, such samples are often statistically quite acceptable. The reason is that once N is so large that the sampling fraction n/N becomes negligible in the formula for $Var(\bar{y})$, then only the absolute value of n is important. From this point of view, it does not matter much whether the sampled population is a city of size 100,000 or the whole nation of 200,000,000 because the variance of a sample mean of size $n = 1000$ is essentially $\sigma^2/1000$ in both cases.

A small hedge is necessary. Sometimes, in larger populations there is a tendency for σ^2 to be larger, even for the same measurement. For example, in a study of incomes covering the whole State of New Jersey, the range of incomes will be at least as large as in a study of a single city within the state. *Sometimes,* this implies a larger variance in the larger target population. However in practice, this increase is rarely anywhere near in proportion to the increase in the population size, and may not exist at all. Usually, small samples can be used quite effectively to yield good estimates for large populations.

In summary, here are some points to keep in mind about sample size estimation.

(1) A formula for estimating n can be obtained, but (a) it depends upon the unknown shape of the sampling distribution *and* its sampling variance, and (b) it is a formula for *one* measurement. How to combine many measurements is largely subjective.

(2) Data errors and selection biases (deck stacking) can make the nominal precision completely misleading, even if all other problems seem to be well in hand.

(3) Selection of a sample size requires the specification of estimators and sampling designs. Consequently, it is first necessary to consider which designs and estimators have the best chance of being efficient.

Exercises

13.4.1 For the population of United States Corporations given in Exercise 11.2.2 make a graph analogous to that in Figure 13.2 showing how the variance decreases as a function of sample size. To complete this exercise you will have to make some decisions; be sure to justify each of them.

13.4.2 You are given the following information on on a population of four strata.

i	M_i	$\bar{Y}_i$	S_i
1	1500	8.5	3.7
2	1500	4.3	1.9
3	1500	8.9	2.8
4	1500	12.6	4.3

(a) Allocate a sample of size 150 to the four strata by proportional allocation and then by optimum allocation.
(b) Calculate the variance of the stratified estimator in the case of proportional allocation and in the case of optimum allocation. Ignore finite population corrections.

(c) Calculate the proportional reduction in variance of optimum allocation over proportional allocation.

(d) Calculate the sampling variance of the sample mean if a sample of 150 is drawn by completely random sampling, that is, by ignoring the strata.

(e) Assuming optimum allocation estimate the sample size required such that the difference $|\bar{y}_{st} - \bar{Y}|$ is less than 1.2 with 0.95 probability. What assumptions do you need?

(f) Answer the same question as part (e) assuming that completely random sampling is used.

13.4.3   Sometimes an analyst will draw "a 5% sample." Discuss why such a relative formulation of sample size is not adequate to estimate the variance of an estimator.

13.4.4   Ten observations were obtained from a large population and the values $\bar{y} = 81.9$ and $s = 450$ were calculated. What statements can be made about the population mean? What assumptions are involved in making these statements?

IMPORTANT NEW IDEAS

estimation of sample size

required precision

permissible risk of failure

CHAPTER 14

WHAT WAS THAT AGAIN?

14.1 Sampling Distributions

The target population is the set of units under study. Examples are the entire United States population, a day's production of light bulbs and, of course, United States taxpayers. In practice, little, or at least not enough, is known about the target population. This is the reason for a sampling study. Populations have many features which may be of interest. A number of these features such as the middle of the population and its spread have been discussed in this book. But, particular emphasis has been placed on estimating the true population mean, $\overline{Y}$, and measuring the accuracy of this estimate.

Our goal has been to develop an *understanding* of sampling design and estimation, and by now it should be clear that the sampling distribution plays the central role. The sampling distribution of an estimator (such as the sample mean $\overline{y}$) is the set of all possible values of that estimator that can arise in the sampling process. In practice, only *one* value of the estimator is usually known, so it may appear that little can ever be learned about any sampling distribution. Fortunately, as we have seen, such a conclusion is incorrect.

Statisticians have proved a series of mathematical results which connect the target population with both the sampling distribution and the sample. These connections permit surprisingly strong statements to be made from only a single sample. The first of these results connects the expectation of an estimator with the target population. In the case of the sample mean, $E(\overline{y}) = \overline{Y}$, which tells us that the mean of the sampling distribution is the same as the mean of the target population. An estimator with this property is said to be unbiased; see Chapter 5.

The next essential connection is between the variance of the sampling distribution and the variance of the population. For example, in sampling with equal probability without replacement, (Chapter 5),

$$Var(\overline{y}) = E(\overline{y} - \overline{Y})^2 = (1 - \frac{n}{N}) \frac{S^2}{n} .$$

229

In this equation, $Var(\bar{y})$ is in terms of S^2, the dispersion of the population, and the known values of n and N. This formula is very handy because it means that, if S^2 can be estimated from a single sample, $Var(\bar{y})$ can also be estimated from a single sample. Fortunately, S^2 can indeed be estimated from a single sample, provided that the sample has at least two observations. As we have shown, $Es^2 = S^2$, so that if s^2 is substituted for S^2 in $Var(\bar{y})$, then

$$var(\bar{y}) = (1 - \frac{n}{N}) \frac{s^2}{n}$$

is an unbiased estimator of $Var(\bar{y})$. This estimate can be obtained even though only *one* sample is actually obtained from the entire sampling distribution! But pleasing as this result is there is more.

We saw in Chapters 7 and 8 that, if really informative statements are to be made, confidence intervals are necessary. But we also saw that confidence intervals not only need the variance connection described in the preceding paragraph, but also need the *complete* sampling distribution. By any reasonable expectation this requirement should be a complete block to the development of confidence intervals. But thanks to the amazing central limit theorem, it is not.

The central limit theorem says that as the sample size gets larger, the *shape* of the sampling distribution of the mean becomes that of the bell-shaped normal distribution. Furthermore, this is true for almost all sampled populations. For populations that are very non-normal, somewhat larger sample sizes are needed, but the result still holds; the bell shape relentlessly appears. The central limit theorem seems to work pretty well, even if the sample size is only in the range of 30, which is smaller than the size of most socioeconomic surveys.

In summary, statistical methods permit a great deal more to be inferred from a single sample than we had any reason to expect, or hope for, before we began the study of these methods.

14.2 Variance Reduction

Intuitively, a 95% confidence interval of 10.1 to 12.9 (say) appears to be more useful in practice than the longer 95% confidence interval of 7.1 to 14.2. We also know that the length of confidence intervals depends directly upon the variance of the sampling distribution. Consequently, efficient design of sampling studies comes down to a search for methods which have sampling distributions with smaller variance.

What are the methods that can be used to obtain smaller variances? Perhaps the selection procedure is the most important. If carefully coordinated with the grouping of the population, natural or otherwise, good selection procedures can realize very large reductions in variance. On the other hand, failure to pay attention to this coordination can hurt badly. It is a two-edged sword. However, the magnitude of the potential gains implies that the method of sample selection should be your first choice method to reduce variance.

The next factor is the choice of estimator. Ratio and regression estimators nearly always have smaller variances than the simple sample mean. This reduction in variance depends upon the availability of auxiliary measurements which are correlated with the measurements under study. Ratio and regression estimators are generally biased, but with moderate attention these biases can be held to a practically insignificant level. On the other hand, ranking the choice of estimator second in importance behind the selection procedure implies sticking to the more or less standard estimators. It is indeed possible to construct bad estimators.

Increasing the sample size is the third most important method of variance reduction. It ranks behind both selection procedures and the choice of estimator. Unfortunately, much statistical literature gives the impression that increasing sample size is the *only* way to reduce variance. Nevertheless, it ranks third because increasing the sample size faces quickly diminishing returns, and can be very expensive both in terms of money and data accuracy. All of this is not to say that sample size questions are not important, they are, but as a method of variance reduction look to both the method of sample selection and the choice of estimator for the biggest gains in efficiency.

14.3 Types of Bias

Bias has been discussed at different places in the text, but the topic is important enough to warrant this summary. All biases have the same effect in that they direct the sample estimate away from the true value. But as bias can arise for different reasons, it is helpful to classify three different types.

(1) Technical Biases. The biases most often discussed by statisticians are of this type. They are a result of the functional form of the estimator not averaging over all possible samples to the true population value. Ratio and regression estimators are generally technically biased. But if standard estimators are used and some attention is given to the known technical bias characteristics of these estimators, this bias source should not present too much difficulty in practice.

(2) Measurement Biases. The sky is the limit on how much difficulty can arise from measurement inaccuracies. The number of ways in which these errors can enter is very large and so is the magnitude of their effect. Measurement devices, whether human or mechanical, can go awry, and in addition, errors can be introduced at the data processing stage. Books have been written on just parts of this problem. But here, it must suffice to say that if the most careful controls are not exerted to ensure accurate data, you almost certainly will wind up being misled. Unfortunately, you may also be misled even if controls are exerted. The reason is that measurement errors are often unique to the particular survey and come as a surprise, in spite of all prior expertise. Who knows how many measurement errors are never found at all!

(3) Selection Biases. These are biases that result from distortions in the selection procedure and they may be the most insidious of all because even the most careful examination of sample data will *not* necessarily reveal their presence. The deck may be severely stacked even though measurement errors have been controlled. It is a case of bad data looking good. Sometimes outside data can be used to discover selection biases and, in these cases, it may be possible to make corrective adjustments. But unfortunately this possibility is not always helpful. There are two reasons. First, there may be no suspicion of a bias in the sample and so no correction is sought. Second, even if a suspicion exists, appropriate exogenous data may not be available to make the adjustment. It is far better to be very careful about the operational aspects of the sample procedure, *in advance.*

Nonresponse is the failure to get observations on some sample units. The occurrence of nonresponse may be a sign of selection bias because there may be a correlation between the ability to get a response and the measurement itself. For example, families with no children are less likely to be found at home than families with children. Furthermore, families with children are different from families without children. For example, their patterns of income expenditure are very different. Consequently, surveys with nonresponse run the risk of under-representing childless families and over-representing the rest. There is a lot of evidence that this does indeed occur.

For the preceding reasons, the correction of samples with selection bias is often very difficult, and is frequently impossible. In many cases, it is only possible to guess as to the characteristics of the missing observations. What you can't measure, you don't know.

CHAPTER 15

HOW TO

15.1 Design Your Survey

In this chapter, a sequence of steps for the actual design of a survey is presented. The sequence has a logic to it, but it is important to realize that the steps are not independent. Decisions made at one step will, to some extent, affect all the others. As a result, the planning of a survey can resemble a merry-go-round rather than a sequence of logical steps. Nevertheless, a discussion of steps to follow in the planning of a survey is worthwhile. Keep in mind that the steps are not hard and fast rules. Understanding is what is important.

The steps are the following:

(1) Avoid an Early Discussion of Sample Size

(2) Formulate the General Goals and Uses of the Survey

(3) Specify The Frame

(4) List The Measurements

(5) Discuss The Sampling

      (a) Special treatment for special cases
      (b) Sample like (unalike) groups lightly (heavily)
      (c) Create artificial groups
      (d) Spread the sample out
      (e) For unequally sized groups use probability proportional to size

(6) Make the Final Sample Decisions

(7) Choose the Estimators

These seven steps are discussed in order below.

(1) Avoid an Early Discussion of Sample Size

Avoid early discussions of sample size. The only exceptions occur when information from previous studies is available or when a pilot study is a possibility. Barring this, sample size cannot be sensibly

233

discussed at the outset of planning.

The temptation to break this rule is strong. Financial supporters are interested in cost, and cost is heavily dependent upon sample size. Unfortunately, the role of clever sampling design and estimators is not clear to many clients, who apparently would be satisfied if they were simply told to "take $n = 839$." Nevertheless, it is usually unwise to begin with a discussion of sample size.

(2) Formulate the General Goals and Uses of the Survey

Begin with a complete discussion of the survey goals. Is the objective of a marketing study of solar heating to determine potential revenue or public acceptability? What is acceptability? What is the target population? Does it include large companies and wealthy individuals? Will the study be confined to a specific geographic region and/or certain sociological groups? It is important that as many of these questions as possible *be answered at the outset.*

In the discussion of goals, the possible subsequent analysis of subgroups of the population is frequently overlooked. To illustrate, a sample of an entire city may later be used to estimate income for black persons in the core area of that same city. Some of these subpopulations can be anticipated before the sample is drawn. Of course, such advance recognition will give any subsequent analysis the best chance of success. At the very least, advance consideration should prevent the subgroup sample size from being grossly inadequate. To help anticipate which subpopulations may eventually be studied, try to identify the various organizations that will have access to the data. Consideration of the functions of these organizations can help greatly to project the future uses of the survey data.

As an operational suggestion write the general survey goals down for future reference; they have a way of being forgotten and distorted. Sometimes, this distortion occurs even before the data gathering begins.

(3) Specify The Frame

Potential target populations must be discussed in a general way along with the goals of the survey in step (2). But soon after the general discussion, a detailed frame must be formulated. Recall from Chapter 3 that frame formulation requires clear and precise definitions of the population units, otherwise substantial difficulties may occur later in the study.

The necessary definitions are not always easy to develop. For example, if a household study is involved, what is meant by a household? Do two single women living together constitute a household? What about 10 single women living together? What if at the time of the survey they are only living together temporarily? What is temporary? Also, is the family with a winter and summer home to be included if the area sample selects the winter home in the summer while they are away? Such questions are not answered in this book, but they must be answered for each survey if the study is to have a reasonable chance of success.

Once the detailed definitions are made, the relationship between the target population and the frame should be reviewed because operational difficulties that become clear while formulating these definitions may result in a shifting of the target population. For example, difficulty in defining households has caused some analysts to shift to a target population derived from local telephone books, that is, telephone numbers are sampled instead of households.

Finally, recall from Chapter 3 that natural groupings of the population units are usually included with the frame. This specification is conveniently done at this same time, and it is very useful later when the sampling methods are specified.

(4) List The Measurements

First make sure that the *right* questions are asked, relative to the goals of the survey. If the marketing study of solar heating has potential profitability as a goal, a question about the likelihood of purchase is confusing if the person who is answering the question and the person who will pay the bill are not distinguished. Also, answers to the question "What was your income last year?" may refer to family income or to the respondent's income. The question is not precisely enough worded. It is very important that the right questions are asked.

Second, after the right question is asked, make sure that the right answer (measurement) is obtained. There are many pitfalls here because a correct measurement depends upon both the sample unit and the measuring device, either or both of which can create measurement problems. Perhaps, the biggest problems come about when people are involved. Estimation of the number of abortions may be a clearly stated goal, but how is the right answer to be obtained? It would be unreasonable to expect this question to be answered honestly, even if moral guarantees of privacy are given. One possible way around this difficulty was given in Section 6.3. As a further example, how can an interviewer obtain a correct value of a family farm income if the interviewer himself does not understand the definition of farm income? In such a case, the measuring mechanism (the interviewer) will not function properly.

It is important that the measurements received really are the ones anticipated. If at all possible, pretest the system under realistic conditions. This does not just mean checking that numbers are rolling into the data processing system. They must also be correct. Be particularly suspicious if people are involved -- otherwise the questionnaires may be filled out in the leisure of the interviewer's hotel room!

As a third point, make sure that *all* relevant measurements are made. Without advance planning, even important variables are easily omitted. A nice example of this seems to keep recurring in work sampling. Work sampling is a specialized area of sampling which is designed to estimate the *proportion* of time spent on various categories of work, in (say) a business office. But if office efficiency is one of the study goals (as it usually is), then more than work sampling data needs to be gathered. The reason is that the proportion of time spent on a particular work category has an ambiguous interpretation. Such a proportion can change either if the work load changes, *or,* if the average time per repetition changes. So, if one business office spends 50% of its time handling mail orders, it is not necessarily more efficient than another office which spends 60%, simply because the second office may handle more of these orders. In this situation counts of the *number* of orders should be taken, in addition to the estimation of proportions. In summary, make sure that all relevant data are gathered. While not necessarily as disastrous as gathering the wrong data, the error of omission can be both embarrassing and costly.

Finally (fourth), review what will be done with each measurement and the accuracy that is needed for this measurement to be practically useful. If counts are made of the number of children per family, is the mean per family to be estimated? Next, review how this mean relates to the overall survey goals and specify the useful accuracy. It is of no practical use to have an estimated mean number of children per family which is accurate only to within three per family! So, for each estimate, state reasonable and useful accuracy expectations. These expectations will be needed, when discussion of the sample size really does arise.

(5) Discuss The Sampling

At this stage an approximate "working" sample size is useful. The best way to pick this initial working sample size is on the basis of cost or other operational factors such as time available. This is not the final sample size but rather is a working sample size that can be used to discuss possible sampling procedures. The reason for this approximation is that the sampling design depends upon the sample size available, and therefore, specifying a design is very difficult without a "working" sample size. The reverse is also true, that is, statistical estimation of sample size is not possible until the sampling design has been formulated. The reason for this circularity is that the variance formulas, upon which the sample size formulas depend, in turn depend upon the sampling design.

At this point, apply the lessons learned in Chapter 11, "The Clever Use of Groups," and decide how the sampling will actually be done. At the risk of being overly simplistic, these lessons have been summarized into the rules of thumb below.

(a) Special treatment for special cases

The separate consideration of unusual population units is frequently a good idea for both statistical and nonstatistical reasons. The example of the large taxpayer in the IRS study is not misleading. There are many similar situations. In marketing studies, large customers are often included in the sample with certainty, and the same is true of the very large United States cities in national samples.

Sampling rule number one says give special treatment to special cases.

(b) Sample like (unalike) groups lightly (heavily)

In the subsection title, this rule of thumb may appear to be two rules, but it is actually just one. It happens that the idea is more easily understood by consideration of the two extreme cases, just as was done in the children's age example of Chapter 11.

First sample like groups lightly. In Chapter 11, the numbers 3, 3, 3, 6, 6, 6, 9, 9, 9 were put into three groups, $\{3,6,9\}$ $\{3,6,9\}$, $\{3,6,9\}$. These are alike groups, in fact, they are *exactly* alike, and we saw that sampling them lightly (very lightly because only *one* group was selected) achieved a zero variance. The selection of a second group would give *no* additional information at all.

Next, sample unalike groups heavily. When the children were put into groups $\{3,3,3\}$, $\{6,6,6\}$, $\{9,9,9\}$, zero variance was achieved by selecting one unit from each of the three groups. These groups are unalike and they were sampled heavily by taking observations from *every* group. It is not possible to sample more heavily than by

observing every group. So this part of the rule tells us to sample unalike groups more heavily.

In practice, such extreme groups don't occur, but the objective is
clear nevertheless. Groups that seem to be homogeneous can be sampled lightly while heterogeneous groups should be sampled more
heavily. Thoughtful application of this rule can achieve very
worthwhile variance reductions.

(c) Create artificial groups

In the chapter on ratio and regression estimators, the creation of
useful, but artificial, auxiliary measurements was discussed. To illustrate the point, "eye estimates" were discussed in the sampling of an
orange grove. Similarly, it is sometimes useful to create artificial
groups. An example once implemented by R. J. Jessen, is an excellent
illustration.

In a survey to estimate the number of ears of corn remaining in
fields *after* the corn had been harvested, rows of corn within the fields
were proposed as sampling units. These rows were known to contain
anywhere from 0 to 12 residual ears of corn. Consequently, if rows
were used as sampling units, the 0 to 12 variability would be reflected
in the variability of the estimator. The clever part of the study is that
the rows were not used as sampling units, but rather, some efficient
new units were created.

The new units were formed by sending a worker *quickly* along the
rows placing a stake everytime *approximately* three ears of corn were
observed. As a result, these new units were of variable row length.
But most importantly, they usually contained from 2 to 5 ears of corn.
Consequently, the variance among these new units was substantially
smaller than the variance would have been, had the complete rows
been used as sampling units. Specifically, the range of observations for
the new units was about $5 - 2 = 3$, and the range of observations on
rows was $12 - 0 = 12$. Since the relationship between the range and
standard deviation can be approximated by $R = 6\sigma$, (see Chapter 13),
the component of variance for the new sampling units was approximately $12/3 = 1/4$ of the variance among rows. This neat procedure
reduced the variance of the estimators substantially. There are few single procedures that will reduce variance by as much as clever definition
of the sampling units. Certainly, it would take a large increase in sample size to do as well. Finally, and most importantly, notice that this
gain in efficiency cost very little - only the initial staking out of the
units. In many cases, creation of homogeneous sampling units can be
carried out without any cost at all.

(d) Spread the sample out

It is usually best to spread the sample out as far as possible over the target population. To illustrate, sample 10% of the households in *all* city blocks, rather than all of the households in 10% of the blocks. The reason is that the "between block" component of total variance is usually larger than the component associated with households. Consequently, a more effective variance reduction is usually achieved by allocating effort to the groups that are higher in the hierarchal organization of the target population.

Operationally, start by a 100% sampling of the largest groups. Of course, this can only be carried out if the working sample is large enough to cover all of these groups. In a sample of all United States urban dwellers, this rule suggests allocating some sampling units to every state rather than randomly selecting some of the states. After that, the same rule of thumb should be implemented with the next units in the hierarchy. So, in the United States urban sample, if the next units are cities, the rule suggests selecting some people from every city. Again, this implies that the sample size is large enough to cover all cities.

Ultimately, this procedure will be impossible to implement because the remaining sample size will be smaller than the number of groups. In the sample of urbanites, if the next grouping is city blocks within each city, the rule will almost certainly break down because there are hundreds of thousands of city blocks in the country, and few samples are anywhere near that large.

So in summary, in a general purpose survey spread the sample over the population as far as possible. At the same time, keep in mind that an interest in prespecified subpopulations may be in conflict with this particular rule of thumb. Unfortunately, rules of thumb to handle this conflict are difficult to envisage.

(e) For unequally sized groups use probability proportional to size

Recall from Chapter 11 that the selection of unequally sized groups with equal probabilities can result in a disastrously large variance. Consequently, after deciding which groups will be sampled, as opposed to completely observed, (Rule d), determine if these groups are of equal or unequal size. If they are of unequal sizes, select them with probability proportional to size.

In the preceding United States urban sample, we suggested that sampling almost certainly must begin at the city block level. If the ratio of largest block to smallest block is larger than about 1.25, plan to use unequal chances of selection.

In practice, unequal probability selection of blocks complicates the sample analysis. To avoid this many practitioners construct city blocks that are nearly equal in size. However, since it is not always possible to manipulate population groupings, keep the rule in mind; unequally sized groups should be selected with probability proportional to size.

(6) Make the Final Sample Decisions

At this point, some specific decisions about sample allocation, sample size and nonresponse are required. First spread the working sample *proportionally* over the groups. Proportional allocation results in a self weighting design and has good variance properties for most measurements. To do this in the sample of United States urban dwellers, allocate the sample to the states in the same proportions as the total number of urban dwellers in the states are to the total population. So if 10% of the total number of urban dwellers in the United States live in New York State, allocate 10% of the sample to New York State. Repeat this at each of the various levels in the population hierarchy.

At this point, the sampling design and a sample allocation have been specified, based on a working sample size. This means that statistical variance formulas can be calculated[1] and compared with the accuracy requirements specified in the survey goals. Of course they won't agree. For some items the estimated statistical accuracy will meet the specified goal; for others it will not. Some compromise is necessary. If it looks as if important items will be estimated with too little accuracy, a request for a larger sample size may well be in order.

Finally, the discussion of any sample must include the possibility of nonresponse. Some observations that should be made will not be made. This can happen whether the measuring process is human or mechanical. The result of nonresponse can be severe bias. If observations are missed randomly, no bias results; but unfortunately, random omission is probably rare. This seems to be true whether the measuring devices are human or mechanical. Mechanical devices do go awry, particularly at the limits of their measurement capability, but the nonresponse problem is particularly acute when humans are involved. For

1. These variance formulas can be calculated, subject to the restrictions discussed in Chapter 13. For example, some prior information must be available on the magnitude of the different components of variance.

example, it is generally agreed that in urban areas there is a hard core of unobservable,[2] unemployed persons. If the rate of unemployment in this group is large, say 20%, the result of this nonresponse will be a downward bias in the estimated overall rates of unemployment.

Nonresponse problems are serious, but the topic is much too broad and complicated to discuss in this book. All that really can be done here is to encourage careful attention to it, *even if* nonresponse is a relatively small percentage (5%) of the data.

(7) Choose the Estimators

The estimators discussed in this book are the ones used most of the time in descriptive surveys. While the properties of these estimators are by no means completely described here, the nature of their use certainly has. So it should be possible for a reader to make a reasonably good choice based almost solely on this book. Confidence intervals are often substantially reduced in length by the judicious choice of an estimator.

Before the survey, specify the anticipated choice of estimator for each measurement in the survey, and review how each relates to the overall survey goals including the accuracy goals. It may be very difficult at this stage to figure how much reduction in variance a particular estimator may bring, because this reduction depends upon factors not usually known before the survey. For example, the reduction in variance due to use of a ratio estimator depends upon correlations which are not usually known in advance of the actual survey. One result is that, in the planning stage, survey statisticians often ignore the reductions in variance possible from a good choice of estimator. This potential gain is an ace-in-the-hole in the event the natural variability in the population is greater than expected. And indeed, the 1/2 to 1/3 reduction in variance that use of ratio and regression estimators often gives, does make a very nice cushion.

15.2 Criticize Another Survey

Intelligent criticism of other surveys is important. Occasionally, there is a need to do this professionally and in detail, but in day-to-day life such criticism and appraisal must be made frequently.

Unfortunately, the widespread use of surveys seems to be paralleled by misleading and partially informative statements. To say that "2

---

2. They refuse to cooperate with interviewers even if found.

out of 3 doctors recommend 'A' for headaches" is not necessarily misleading, but it is certainly only partially informative. Clearly, the statement is not based on all doctors. If so, is there a target population? How was it sampled? How were the questions asked? How big was the sample? What about confidence limits? The unanswered questions certainly leave a great deal of room for speculation.

The advertising industry is not alone in making misleading and partially informative statements. They can be found in many fields including engineering, medicine, and journalism. In the last field, it is noteworthy that the New York Times now routinely includes a special section describing the sample design of their surveys. The Times is to be commended for this, but unfortunately, such action is rare. In most popular survey reports, an unfortunate amount of speculation about the results is possible.

Perhaps the best way to systematically appraise another survey is to review the points discussed in Section 15.1. The procedures that require attention in your own survey are the same ones that someone else overlooks. However, to give the criticism a different point of view, some points to look for are listed below.

(1) Focus on the Measurements

To do this ask yourself a number of questions.

First, was the wrong question asked? If a survey asks for age of the respondent and really wants age of the head of the household, the wrong question was asked. Always check to see if the right measurements were taken. Do this every time you are confronted with sample results and you will be amazed how often that the right question was not asked.

Next, was a wrong answer obtained to the right question? In personally sensitive interviews, it is quite easy to ask exactly the right question but still get wrong answers. Such errors lead to poor estimates. For this point, recall that subsampling the data, and rechecking it, is a very useful procedure. Never assume that measurements are made properly.

Was the wrong interpretation put on the result? Are there alternative explanations? It is often demonstrated that joggers have better cardiovascular systems than nonjoggers. The customary conclusion is that to live a long time one should get out and jog. But is it possible that cause and effect are mixed, and that joggers jog because they have better cardiovascular systems in the first place? Similarly, do smokers smoke because they are genetically different from nonsmokers, and this genetic difference is related to cancer? Cause and effect can never be assigned by statistics alone, and there are often alternative explanations for statistical relationships.

Finally, were all the measurements made that are necessary for the conclusion? Adams and Murray[3] describe a nutritional survey in which certain biochemical tests relating to the good health of individuals were made. But, no parallel clinical observations were made. So, no matter what was done correctly in the study, the conclusions are suspect, because critics feel that the biochemical measurements *alone* did not necessarily indicate good health. Since the assessment of good health was the overall goal of the survey, the missing measurements were a major oversight. Clearly, not all the necessary measurements were taken.

(2) Selection Biases

Selection biases can have very large effects and, at the same time, be very difficult to detect. Data can be measured correctly, analyzed correctly, but be very badly biased. In fact, if a sample were selected to deliberately mislead, selection bias would be the way to do it. Someone else can recheck the measurements and redo the analysis and still be misled by selection bias. In this sense (see Chapter 10) samples with selection biases are *dishonest* samples.

Detection of a selection bias can be very difficult, but in the evaluation of someone else's survey, first try to find out if there is a coverage problem. To do this, compare the target population and the sampled population. Usually, the target population will be made more or less clear in the description of conclusions, but the same may *not* be true for the sampled population. The sampled population may not be made clear at all. To infer what the sampled population is, study the actual sampling operation. This can reveal a great deal. A sampling of persons in hospitals is not close enough to a target population of the whole United States population; there is a major coverage problem. On the other hand, in many areas of the country, sampling "telephone households" is probably satisfactorily close to a target population of all households. A great deal of judgement is often required on this point, but you will be surprised how often study of the sampling operation reveals that the sampled population is unwittingly different from the

3. Adams, Ruth and Murray, Frank; Vitamin E, Wonder Worker of the 70's, Larchmont Press, 1971, p. 116.

target population.

After consideration of possible coverage problems, review the sampling procedures again, to determine if the sample selection favors any special groups. For example, if a politician decides to count favorable and unfavorable telephone calls received after a speech, he could arrange to have a switchboard operator steer favorable calls to a group of operators which is clearly large enough to take the calls with no waiting. On the other hand, unfavorable calls could be directed towards an undermanned staff, making it difficult to get through. Results in his favor are greatly increased. In fact, the politician could then even show his character by passing the basic data tallies to his opposition for totalling and analysis. Then (provided the opposition is honest), the result is almost a foregone conclusion. His speech will likely receive a favorable assessment because favorable calls have a higher chance of being counted than unfavorable ones.

Finally, if possible, examine the sample and compare it with available external data. Are males and females proportionally represented? Is the distribution of family size the same as in the population at large? Unfortunately, the external data necessary for such comparisons do not always exist. The politician, who directs his telephone calls in the way described in the preceding paragraph, cannot be caught if he is smart enough to pass along only the basic tally to the opposition (the public?), with no information about how the tally counts were obtained. He could, of course, honestly point out that *all calls received* were counted.

(3) How Variable Can the Results Be?

What is the possible range of the quoted results? Could the "2 out of 3 doctors" just as easily be 1 of 3 or even 0 of 3? Such information may or may not be given along with the results, but some idea about variability is necessary for proper assessment of conclusions. Confidence intervals are very informative, but usually they are not given.

As we have seen, confidence limits depend upon a number of factors including, the sample size, the sampling procedure, the estimator, and the natural variation of the population. So getting a confidence interval really requires consideration of each factor. An approximation can be made by the range trick described in Chapter 13. That is, divide the empirical range (the largest in your experience observation minus the smallest) by $6\sqrt{n}$. Then use this as an estimate of standard error to develop a confidence interval. This procedure is crude, but will give an approximate idea of the precision of the survey that has fallen to you for evaluation. Usually such a variance estimate (and the resultant confidence interval) is conservative because it takes no account of possible variance reductions achieved by clever design and efficient estimators. On the other hand, we say *usually* conservative because poor sample design and poor estimators may result in variances that are *larger* than simple random sampling.

INDEX

A

Absolute size of sample, 226
Accuracy of samples, 45-47
Actual and reported incomes, 30-32, 185-187
Adams, Ruth, 243
Admissions to university, 3-5
Aggregated data,
  misleading effects of, 4
Aging population, illustration, 12
Airlines
  division of revenue, 48
Allocation of samples, 237, 239-240
Allocation in stratified sampling,
  proportional, 165-6, 173-174
  optimum, 165-166
Analogy estimators, 184
Approximation,
  central limit, 103, 216, 223
  to populations, 37-41
  to the range, 224, 238, 244
Artificial groups, 28
Assets of corporations, 143
Attributes,
  see proportions
Auxiliary measurements, 183-184, 238
Average, of a population, 31-33
  of a mean, 53

B

Bases, data, 173
Bell System business offices, 11-12
Best confidence intervals, 81
Betting odds, 127-128
Bias:
  in the club meeting, 11
  in common language, 68
  effect on confidence intervals, 101-103
  in estimators, 68

first-month, 22
and the mean square error, 99-100
from measurement errors, 69-72, 232
in panel surveys, 12-14
in ratio estimation, 194-195
in regression estimation, 200-201
selection, 68, 127-129, 174-175, 232, 244-245
by special groups, 1
in stratified sampling, 163-164
technical, 68, 100, 209-210, 231
in television ratings, 21
in unemployment, 23-24
Binary numbers, 8
Business offices, Bell System, 11-12

C

Call backs, 131
Cancer, lung, 18
  uterine, 16-17
Census, definition, 44
Census Bureau, 16, 21, 46, 71
Central limit
  approximation, 84-87, 103, 216, 223
  illustration, 84-87
  theorem, 103, 216, 223, 230
Chance of selection, 8, 76-77
  equal, 52
  overall, 57, 164
  proportional to size, 160, 239-240
  unequal, 105
Chatham Township, N.J., 156
Children, disappearing, 14-15
  grouping of, 144-146
Choice of estimators, 241
Club meeting date, bias in, 11
Cluster sampling, 146, 148
Collapsed strata, 168
Combinatorial notation, 53
Conditional unbiasedness, 206

"First-month" bias, 22
Fixed panel surveys, 12-13
    see also sampling through time
Fixed slope regression estimator, 200-201
Forecasts of rate of return,
    of 500 U.S. Corporations, 37-41
    skewness of, 41, 84-87
Frame, 28, 234-235
Freeny, Anne, 20
Frequencies, relative,
    see relative frequencies

G

Gabbe, J., 70
Gallup poll predictions 1948, 97-98
Gamma approximation to forecast errors,
    39-40
Goals of surveys, 234
Grouping,
    children by age, 144-146
    gains and losses by, 144-146
    misuse of, 146
    population, 28, 231, 238
    for zero variance, 145
Groups,
    artificial, 28
    natural, 28
    primary, 28, 148, 167
    secondary, 28, 148, 167
    tertiary, 28, 148, 167
    unequally sized, 153-162, 239

H

Hartley-Ross unbiased ratio estimator,
    195
Harvard Business School, 20
Histogram
    definition, 38-41
    of forecast errors, 38-41
    of taxpayer population, 211-214

I

Income, actual and reported, 30-32, 185-

187
    distribution of, 32
    of families, 41-42
    taxpayer population, 30-31
Interaction of measurement and meas-
    urement process, 12
Interpretation of data, 2, 16
Intervals, confidence,
    see confidence intervals
Inverse weighting, 120-121, 126, 151-152,
    157, 160

J

Jessen, R. J., 238
Joggers, 19-20
Judgement samples, 47

K

Kinsey report, 17-18

L

Large taxpayer,
    special treatment of, 50-51, 137-42
Limits, confidence, 78-81
    see also confidence limits
Lindsay, John, 23
Location, measures of, 31-33
Lottery, 1970 draft, 5-7
    1971 draft, 9-10
Lung cancer, 18

M

Mail survey, 131-132
Mallows, C. L., 24
Mean,
    of a normal distribution, 88
    of a population, 31-33
    of a sample, 53
    sampling distribution of, 55-57
Mean square error,
    definition, 99-100

Rotation sampling
   illustration, 12-13
   see also sampling through time

                    S

Samples,
   accuracy of, 45-47
   allocation of, 165-166, 173-174, 208,
      237, 239-240
   cost of, 45
   distribution of, 65-66
   judgement, 47
   mean of, 53
   misleading, 4, 76-77
   necessity of, 44
   ordered, 182
   representativeness of, 2
   retrospective, 18-19
   reweighting of, 15, 129, 202-208
   satisfaction with, 48
   single, 63-65
   size of, see Size of samples
   spread of, 33, 53
   statistical, 44, 47
   variance of, 33, 53
Sample units, 27, 166
   use in variance reduction, 238-239
Sampling, cluster, 146, 148
Sampling, quota, 132
Sampling, restricted, 153, 162
   see also stratified and multistage
      sampling
Sampling, systematic, 105, 132-135
Sampling design in practice, 236-240
Sampling distribution
   discussion of, 229-230
   factors which influence the, 61-62,
      105
   of the sample mean, 55-57, 107-109
   of the sample variance, 55-57
   variance of the, 60-62
Sampling error, 45-46
Sampling through time, 1, 12-15, 25
Sampling with equal probability with
      replacement,
   comparison with sampling without
      replacement, 110-112

description, 62, 105-107
relative frequencies, 109-112
sampling distribution of $\bar{y}$, 107-109
Sampling with equal probability without
      replacement
   comparison with sampling with
      replacement, 110-112
   description, 52-53, 64-65, 69-70, 105,
      229-230
Sampling with unequal probability with
      replacement
   bias, 119
   description, 105,
   and the estimator, $\bar{y}_p$, 120
   and grouping of the population, 137
   and probabilities proportional to size,
      122-126
   relative frequencies, 119-121
   sampling distribution of $\bar{y}_p$
   variance of $\bar{y}_p$, 121
   the zero variance case, 122-123
Sampling units, 28, 148, 167
Sampling, work, 236
San Diego weather, 1-3
San Francisco Chronicle and Examiner,
      30
Satisfaction with samples, 48
Secondary sampling units, 28, 148, 167
Selection bias, 68, 127-129, 174-175, 232,
      244-245
Self-weighting designs, 173-174
Sensitivity to outlying observations, 182-4
Sex, 17-18, 73-75
Single samples, 63-65
Size of sample,
   absolute, 226
   and confidence intervals, 214
   diminishing returns from increasing,
      211, 214-215
   discussion, 211-228, 231, 233-234,
      244
   effects of increasing, 84-87
   and estimation difficulties, 227
   a formula for, 214-221
   and guesswork, 223-224
   and multivariate surveys, 224
   and the normal distribution, 230
   prior requirements for, 220-221
   by trial and error, 214-216